Plant Resources for Food, Fuel and Conservation

Plant Resources for Food, Fuel and Conservation

Robert Henry

London • Sterling, VA

First published by Earthscan in the UK and USA in 2010

ISBN: 978-1-84407-721-2

Typeset by 4word Ltd, Bristol
Cover design by Susanne Harris

For a full list of publications please contact:

Earthscan
Dunstan House
14a St Cross Street
London EC1N 8XA, UK
Tel: +44 (0)20 7841 1930
Fax: +44 (0)20 7242 1474
Email: earthinfo@earthscan.co.uk
Web: www.earthscan.co.uk

22883 Quicksilver Drive, Sterling, VA 20166-2012, USA

Earthscan publishes in association with the International Institute for Environment and Development

A catalogue record for this book is available from the British Library

Library of Congress Cataloging-in-Publication Data

Henry, Robert J.
Plant resources for food, fuel, and conservation / Robert Henry.
p. cm.
Includes bibliographical references and index.
ISBN 978-1-84407-721-2 (hardback)
1. Botany, Economic. 2. Energy crops. 3. Agrobiodiversity conservation. I. Title.
SB107.H46 2010
333.95'3–dc22

2009031358

At Earthscan we strive to minimize our environmental impacts and carbon footprint through reducing waste, recycling and offsetting our CO_2 emissions, including those created through publication of this book. For more details of our environmental policy, see www.earthscan.co.uk

This book was printed in the UK by CPI Antony Rowe.
The paper used is FSC certified.

Contents

List of Figures, Tables and Boxes *vii*
Preface *xi*
Acknowledgements *xiii*
List of Acronyms and Abbreviations *xv*

1 Plants for Food, Energy and the Environment 1

2 Plant Resources for Food and Fibre 7

3 Impact of Climate Change on Food and Fibre Production 37

4 Energy Resources 47

5 Plant Resources for Bio-energy and Chemical Feedstock Uses 53

6 Competition between Food and Fuel Production 81

7 Plants, Biodiversity and the Environment 91

8 Impact of Climate Change on Biodiversity 109

9 Competition between Agriculture and Biodiversity Conservation 115

10 Domestication of New Species 125

11 Options for the Future 143

References and Further Reading *163*
Index *175*

List of Figures, Tables and Boxes

Figures

2.1	World cereal production	10
2.2	Wild barley plants are shown growing in a field near Aleppo, Spain	15
2.3	Genetic origin of bread wheat	18
2.4	World population growth	20
2.5	Global food consumption by industrialized and developing countries	21
2.6	Global food consumption by region	22
2.7	Global food consumption – world	22
2.8	The concept of the gene pool	25
2.9	Wild grape and rice relatives	26
3.1	Carbon dioxide concentrations in the atmosphere	38
3.2	Global temperature	40
4.1	Oil consumption – regions	48
4.2	Oil consumption – world	49
5.1	Difference in biomass composition in flowering plants	58
5.2	Interactions between biomass, conversion technology and fuel molecule	62
5.3	Global ethanol production	63
5.4	Ethanol produced from sugarcane on sale in Brazilia	64
5.5	Sugarcane production	68
5.6	Sugarcane	69
5.7	Comparison of conventional cultivation of Eucalypts and growth of a related Melaleuca for annual harvest	73
5.8	Transportation of biomass for biofuel production	76
6.1	Brazil – land use	85
7.1	Relationships between different groups of flowering plants	95
7.2	Comparison of a weed in a new environment and in its native habitat	100
7.3	Biodiversity in cultivation – traditional village gardens near Tsukuba, Japan	102

8.1 The rare *Banksia conferta* 112
9.1 Rainforests 123
10.1 Fertile Crescent – site of early domestication 127
10.2 Ancient grain 128
10.3 Loss of genetic diversity in domestication 132
10.4 Domestication in the Proteaceae 135
10.5 Davidson's Plum 136
10.6 Relationships between higher plants 141
11.1 Mayan ruins on the Yucatan Peninsula, Mexico 147
11.2 Hybrid Eucalypts – Hybrid vigour 149

Tables

1.1 Human uses of higher plants 5
2.1 Flowering plant groups used as foods 8
2.2 Domestication of maize 16
2.3 Rice genus 28
2.4 Rice tribe 28
2.5 Seedbanks for major crop germplasm resources 30
2.6 Consultative Group on International Agricultural Research (CGIAR) Centres 34
4.1 Electricity generation from non-fossil fuel sources 50
5.1 Examples of the composition of plant biomass and predicted ethanol yields 59
5.2 Examples of different stages of development of biofuel technology 62
5.3 Biofuel technologies and products 62
5.4 Examples of plants that have been considered as a source of biomass for biofuels 67
6.1 Examples of arable land estimates 83
6.2 Impact of different levels of biofuel production on food supply (assuming no expansion in agricultural land) 89
9.1 Land requirements to satisfy food and fuel requirements by 2085 – relative to 2005 values 119
9.2 Estimates of yields, conversion efficiencies and areas of land required to replace oil with biofuel in a country consuming around 100GL/year fuel consumption 120
9.3 Water use for electricity and biofuel production from different crops 121
10.1 Domestication of crop plants 130

Boxes

2.1 Cereals 10
2.2 Pulses 12
2.3 Barley: the first plant domesticated 15
2.4 Human selection of rice as an attractive food 17
2.5 DNA banks for conservation and support of plant improvement 31
2.6 The Green Revolution 32
2.7 Improving the folate content of cereals 35
3.1 Scientific and popular views of climate change 39
3.2 Pathways of photosynthesis 41
3.3 Gene diversity in relation to climate 42
3.4 Case study – cereals (rice, wheat, barley and sorghum) in Australia 45
4.1 Approximate conversions for units of energy 49
5.1 Advances in technologies for the analysis of plant carbohydrates 57
5.2 Biomass transportation 76
5.3 Nanotechnology provides greatly improved tools for analysis of plant genes 80
6.1 Arable land in Western Australia 84
7.1 Case study – Queensland, new species 92
7.2 Coastal Fontainea – an example of a critically endangered plant species 94
7.3 What is a species? 97
7.4 Ecosystem services 101
7.5 Lismore Rainforest Botanic Gardens 103
8.1 Impact of climate on wild barley populations 110
8.2 Climate change and *Banksia conferta* 111
10.1 Did plants and animals domesticate humans? 128
10.2 Parallel domestication of plants and animals 129
10.3 The domestication of rice 131
11.1 The case of the Mayan society 147
11.2 Advances in DNA fingerprinting techniques for use in plant improvement 150
11.3 Plant genomics 151
11.4 Research targeting better health and functionality in foods 153

Preface

The capacity of the biosphere to support population growth is being challenged by several emerging issues. Climate change may reduce the productivity of agriculture globally. Growing affluence in major developing countries is creating very strong growth in demand for food. Some have argued that the ability of agricultural production systems to meet this demand may be restricted by the emerging use of crops to produce renewable fuels. The production of biofuels is a response to both the threat of climate change and reducing supplies of affordable oil. These combined demands on land for agricultural production to support strong growth in food and energy consumption threaten new pressures on space for biodiversity conservation. At the same time climate change poses a direct threat to biodiversity. The challenge can be defined as balancing the advantages to biodiversity and food production of any climate change mitigation achieved by biofuel production against any loss of biodiversity and food production resulting from displacement by biofuel crop production. This indicates that we should also aim to ensure that human energy requirements are met as far as possible from parts of plants not useful as food, or plants that can be grown on land not suitable for food production. We need to minimize the land footprint of food and energy production to allow space for biodiversity conservation.

Global analysis of the strategic options available suggests that land, water and other resources use needs to be very carefully managed, especially with the prospects of significant climate change.

Some of the key related global issues are:

- Which of the available plants should be allocated the limited land, water and other resources to produce crops for direct use as human food, pastures for food animals, fibre and energy production?
- How do we resolve competition for these resources between these alternative uses of plants?
- Should we also be growing plants specifically to protect the environment and to conserve biodiversity?
- What plants do we have in cultivation for these applications or can we develop new varieties from wild plants for these purposes?

- How do we balance these needs against nature conservation as an alternative land use option?

This book explores the challenges of selecting from the available plant genetic resources the varieties necessary to meet these new and increasing demands. The expansion of agriculture to satisfy human demand for food and energy will continue to be constrained by the need to support conservation of the biodiversity upon which agricultural production and life itself depends.

Acknowledgements

I thank Kylie Lindner and Linda Hammond for assistance with back up of the manuscript and correspondence with the publisher, and Linda Hammond for assistance in preparation of the figures. Figure 10.2 was provided by James Helm from the photographs of the late Robert Metzger. I thank Tim Hardwick, Earthscan, for assistance with source material and editorial input. I also thank three anonymous reviewers for their constructive criticisms of the manuscript. In an attempt to provide a more first-hand account of the issues discussed in this book, I have, as far as possible, drawn on personal experience and used examples of research in which I have been involved, or plant species from agricultural production systems or wild plant populations with which I am familiar. I thank the many colleagues, friends and family members who have spent the time to introduce me to these plants, their biology and utility.

List of Acronyms and Abbreviations

C 3, C 4	three-carbon compound, etc.
CBOL	Consortium for the Barcode of Life
CGIAR	Consultative Group on International Agricultural Research
CIAT	International Centre for Tropical Agriculture
CIFOR	Centre for International Forestry Research
CIMMYT	International Centre for Wheat and Maize Improvement
CIP	International Potato Research Centre
CO_2	carbon dioxide
DMF	dimethylfuran
DNA	deoxyribonucleic acid
FAO	Food and Agricultural Organization
HMF	hydroxymethylfurfural
ICARDA	International Centre for Agricultural Research in the Dry Areas
ICRAF	World Agroforestry Centre
ICRISAT	International Centre for Agricultural Research in the Semi-Arid Dry Tropics
IEA	International Energy Agency
IFPRI	International Food Policy Research Institute
IRRI	International Rice Research Institute
IUCN	International Union for Conservation of Nature & Natural Resources
LCA	life cycle assessment
NASA	National Aeronautics and Space Administration
PHA	polyhydroxyalkanoate
PLA	polylactic acid
ppm	parts per million
WARDA	Africa Rice Centre
WHO	World Health Organization

1

Plants for Food, Energy and the Environment

Not only are the flowering plants the largest and most successful plant group today but they are of fundamental importance to the life and survival of [humans]. [We] in fact depend on them for major sources of food and sustenance, either directly through agricultural or horticultural crops such as cereals, legumes and fruits, or indirectly through their ability to provide pasture or feed for animals [we] eat. They also provide a source of raw materials for building and shelter, for the manufacture of paper, fabrics and plastics, and for oils, fibres, waxes, spices, herbs, resins, drugs, medicines, tannins, intoxicants, beverages – the list seems endless.

Vernon Heywood, 1978

Background

Plants are a major feature of the landscape over much of the land surface of the earth except for extreme deserts, very high mountains and the Poles where it is too cold. Photosynthetic plants are an essential component of the biosphere, using light energy from the Sun to capture carbon dioxide from the atmosphere, forming the basic organic compounds on which other life forms depend and generating the oxygen in the atmosphere. Modern everyday human life still depends on plants for food as much as our ancestors did. However, many of us, especially those of us living in large cities, are unaware of these links in our daily lives. More than 10,000 years ago humans began the domestication of plants and animals and, in developing agriculture, were able to support large populations that could settle in one place. This process probably happened first in the Fertile Crescent (an area east of the Mediterranean Sea) and then separately in several locations around the world. The development of agriculture was a key factor in humans being able to establish large communities co-located in permanent structures that became our towns and cities. Progressively, the development of human societies has resulted in most people being separated from the daily reality of food production as this task has fallen to an increasingly small and specialized fraction of the population. In industrialized countries, rural populations account for only about 20 per cent of the population (FAO, 2005). We have a general and worrying lack of public appreciation of our continued dependence on domesticated plants (and animals) for food, as most people living in large cities have little opportunity to directly experience the food production systems on which they depend.

Current situation

Plant domestication and the production of large quantities of food in agricultural systems have allowed human populations to continue to expand. This has been possible because of the application of science in the form of plant breeding and agronomy. These technologies have made possible the increases in productivity necessary to keep pace with population growth and agriculture now occupies a large part of the land surface that is suitable for agriculture (the arable land). The human population passed 6 billion around 2000 and is projected to rise to as many as 9 billion by 2050 (FAO, 2005), and agricultural productivity has been dramatically increased to meet the demands of this growing population. However, in very recent times there are signs that we may have reached some limits in our ability to continue to increase food production faster than that required to match population growth or at least the growing demands of human populations. The causes of shortfalls in production are difficult to identify: are we reaching the physical and biological limits of production, are market signals influencing production or have all the benefits of the available technologies been implemented? Sustainable food production requires conservation of the land and water systems needed to support plant growth on the scale required to feed human populations. The availability and cost (economic and energy) of the soil nutrient inputs (fertilizers) required to maintain the nutrient status of the soil are an increasing constraint. The growing economic and environmental cost of energy is making a significant contribution to the cost of producing and distributing these essential inputs to food production. The establishment of agricultural systems that will support the human population sustainably long term remains a challenge. The area of arable land per person is predicted to decline from 0.21ha per person in 1997/99 to 0.16ha per person in 2030 (FAO, 2005).

Recent rising living standards in developing countries have put more pressure on agricultural production systems. A rapid growth in affluence in countries with large populations (e.g. China and India) has increased demand for food beyond that resulting from population growth. Greater per capita demand for food imposes a larger effective environmental and agricultural footprint per person as populations continue to grow. Increased consumption of meat (or animal products) associated with higher incomes is a key factor. Human diets with a higher animal content greatly increase the demand for plant products such as feed grain relative to that of the same human population eating plant products directly. Significant increases in food prices were widespread in 2007/8, fuelling concerns about global food security. If all humans increase their consumption of food to the current levels of developed countries, food demand may be difficult to meet sustainably and biodiversity may be reduced unless new technologies in the form of superior plant varieties or better management become available to allow sustainable production at higher levels.

Human societies do not depend on food alone; they are also consumers of large quantities of energy. Many use plant biomass directly as a source of heating or for cooking. However, fossil fuels such as oil are used for transport and the production of a diverse range of materials (e.g. plastics) that are widely used in modern human societies. The general consensus is that oil is likely to be in short supply in the relatively near future as we deplete world stocks (Hirsch, 2007). The price of oil has spiked sharply recently, partly in response to these perceptions. The burning of fossil fuels has increased carbon dioxide concentrations in the atmosphere. This increase in greenhouse gas concentrations is predicted to result in dangerous levels of global warming. Increasing global temperatures are a new issue that may pose new constraints on agriculture and food production. Major and rapid changes in climate risk widespread extinctions of species of plants, animals and micro-organisms. This situation has created a great interest in developing alternative energy sources to replace oil. The incentives are two-fold: oil is going to run out eventually so scarcity and the associated rising price will drive the search for alternatives; and environmental concerns and constraints imposed by global warming provide another even more compelling incentive. Global warming resulting from the continued use of fossil fuels may directly threaten agriculture and food production.

Plants have been co-opted directly to the production of energy with the recent rapid growth in the conversion of plants to biofuels for use as transport fuel and the burning of plant material to produce electricity. The use of plants to replace oil in the manufacture of products such as plastics is a significant option. This has immediately put more pressure on the agricultural production system. Energy and food crops may be competitors for land, water and other agricultural inputs.

Expansion of human populations has resulted in the displacement of many other species with the rapid extinction of many plants and animals. Large numbers of species are currently endangered and if current trends continue we will see an ongoing rapid loss of biodiversity globally. A major contributor to the environmental impact of humans is the large area of the land surface (especially in favourable environments) that is required for agriculture. The continuing growth of human populations, growth in per capita consumption of food and the additional potential demand for energy crops with declining oil stocks and the threat of global warming – all contribute to a potential acceleration of loss of biodiversity.

Human societies are now dependent for survival upon domesticated plants that are very different in many essential characteristics from the wild plants that were their ancestors. Despite this we also depend upon continued availability of the wild populations as a source of new genetic variation to allow us to adapt our crop plants to environmental variation. New diseases and plant pathogens (biotic stress) and environmental conditions, including short- or long-term climate change (abiotic stress), are a continuing threat to food security. Plant breeders are expected to deliver ongoing food and energy security in an environment that may be increasingly hostile to crop growth, and with the wild genetic resources available for their use in delivering these outcomes under growing threat from development or environmental change. Science is continuing to offer new technologies that have potential to meet these challenges. However, growth in human societies and the resulting demand for food and energy is such that we are required to make major technology advances (e.g. better plant varieties and effective management systems) more and more frequently. Unfortunately this is not just an option, especially if we wish to satisfy human requirements sustainably without major loss of global biodiversity.

Summary of major questions to be addressed for the future

This book aims to define these issues and explore solutions largely from a scientific or technical perspective. Many questions need to be answered or at least raised so that answers can be sought in the future. Can we select more appropriate plants or more efficient plants for food, feed and energy applications? Can we learn from the history of domestication of plants to date? How do we avoid or minimize competition between these uses? Do we need to domesticate completely new species for these new applications? How do we protect biodiversity in all of this? Ultimately, how do we minimize the impact of humans and their agriculture on the global environment and make life on earth more sustainable? What contribution does science and technology need to make to achieve a sustainable future? What types of innovations are required and how feasible are they?

Outline of this book

Human uses of plants (Table 1.1) for food, feed and fibre will be described in Chapter 2. Plant domestication was a major step accelerating the development of large human societies that are interdependent with domesticated plants. The diversity of plants used, their origins and the scale of production necessary to support human populations will be evaluated. The potential of climate change to impact upon our ability to produce the food and other products on which we currently depend will be discussed in Chapter 3. The needs of human societies for energy and the range of sources of energy will be explored in Chapter 4. Chapter 5 addresses the use of plants as a source of energy and specifically the potential of plants to make a major contribution to transport fuel. The resulting potential for competition between food and energy uses of crops will be examined in Chapter 6. The importance of plants in the environment, specifically the contribution of plants to biodiversity, will be explored in Chapter 7. We can replace some of the contributions of natural plant communities to the environment by equivalent volumes of agricultural and forest plantings. However, these do not support the biodiversity of more complex plant communities. Chapter 8 will focus on the continuing competition between the growth of plants for agriculture, forestry and emerging bioenergy applications, and conservation of plants in more diverse natural ecosystems. The loss of biodiversity and the impact of climate change on biodiversity will be examined in Chapter 9. Chapter 10 will explore the potential for domestication of new plant species to better satisfy human need for food and energy while preserving biodiversity. Chapter 11 will provide options for the future and will set out scenarios that could be chosen to deal with the challenges posed by continuing expansion of human

Table 1.1 *Human uses of higher plants*

Category of use	Examples
Food	Cereal, pulses, fruit, vegetables, oilseeds and sugar
Beverage	Wine, beer, tea, coffee
Animal feed	Pastures, fodder
Fibre	Cotton, hemp and paper
Fuel	Firewood
Construction	Housing and furniture
Medicine	Pharmaceutical products, traditional medicines
Ornament	Cut flowers, pot plants, garden plants and turf grass
Industry	Ethanol for fuel, electricity
Environment	Environmental restoration, greenhouse gas sequestration
Other	Perfumes, cosmetics

Source: Henry, 2005a

impact on the global environment. At the end of the book some recommended actions in relation to plant genetic resources will be provided that will support a sustainable future for life on earth.

This book discusses plants in the environment but most of the discussion relates to all organisms, whether plant, animal and micro-organisms. Plants are a useful proxy for all life forms when we consider the impact of human activities and climate change on biodiversity. The diversity of other organisms in the environment will frequently parallel that of the plants which are frequently the largest or most obvious components of the ecosystem. Loss of plant biodiversity directly contributes to a loss of biodiversity of other organisms by reducing the diversity of habitats or micro-environments.

We will begin by considering the use of plants for food, arguably the most fundamental and essential use of plants by humans.

2

Plant Resources for Food and Fibre

The history of the world my sweet is who gets eaten and who gets to eat.

Sweeney Todd, The Demon Barber of Fleet Street (as cited by Lang and Heasman, 2004)

What do we eat? (biological sources)

Human diets are predominantly based upon direct consumption of plants. Animal products are a smaller part of most human diets, but tend to be consumed in greater quantities by more affluent human populations. Animal production is supported by the growth of plants as feed, often consumed by the animal as pasture or fodder or in more intensive production systems as grain. The production of food in animals is relatively inefficient. For example, it takes 3000 litres of water to produce 1kg of rice but 15,000 litres of water to make 1kg of beef, because more than 10kg of feed is needed to produce the 1kg of beef (Millstone and Lang, 2008).

Around 300,000 species of flowering plants have been defined. Relatively few of these species are used as food. However, humans have explored this diversity extensively to find food plants and as a result foods are derived from a genetically diverse range of plant species (Table 2.1). A significant number of species are consumed at least regionally as human foods, but most human food by volume is accounted for by a very small number of plant and animal species. Modern human societies are critically dependent on this small number of species for survival. Despite this remarkable human reliance on these foods, public awareness and as a result international effort devoted to conservation of the plant genetic resources of the major crops on which we depend does not always seem to match the importance of the task.

Food production and consumption has grown to meet the demands of strong population growth resulting from advances in medical science and associated longer lifespans. The total amount of food produced continues to grow largely due to improvements in plant genetics (plant breeding) and

Table 2.1 *Flowering plant groups used as foods*

Nymphaeaceae	**Food** (seeds and rhizomes)
Chloranthaceae	**Beverage** *(Chloranthus officinalis)*
Canellales	**Food** white cinnamon
Piperales	**Food** pepper **Beverage** kava
Laurales	**Food** avocado, cinnamon, bay leaves
Magnoliaies	**Food** nutmeg, custard apple
Alismatales	**Food** *Sagittaria (tubers)*
Asparagales	**Food** onions, garlic, leek, vanilla, asparagus saffron
Dioscoreales	**Food** yams
Liliales	**Beverage** sarsaparilla
Pandanaies	**Food** (starchy fruits)
Arecales	**Food** coconuts, copra, dates, sago, palm oil
Poales	**Food** rice, wheat, maize, barley, sorghum, millet, sugarcane, bamboo, pineapple **Animal feed** pastures
Zingiberales	**Food** banana, ginger, cardamom, tumeric, arrowroot
Ranunculales	**Food** fruits
Proteales	**Food** Macadamia
Caryophyliales	**Food** Ammaranthus
Saxifragales	**Food** grapes, gooseberries, currants
Myrtales	**Food** cloves, lillypilly
Celastrales	**Beverage** Arabia tea
Ericales	**Beverage** tea
Gentianales	**Beverage** coffee
Lamiales	**Food** olives
Solanales	**Food** potato, aubergine, tomato, pepper, sweet potato
Apiales	**Food** carrot, celery, parsley, fennel, dill
Asteraies	**Food** sunflower, lettuce, chicory, Jerusalem artichoke
Dipsacales	**Beverage** elderberry (wine)
Malpighiales	**Food** cassava, passion fruit
Fabales	**Food** peas, beans, peanut (groundnut), soybean **Animal feed** clover, lucerne
Rosales	**Food** fruits (apple, plum, pear, cherry, mulberries, fig, raspberries, strawberries) **Beverage** hops
Cucurbitales	**Food** cucumber, pumpkin, melon
Fagales	**Food** chestnut, walnut, pecan
Brassicales	**Food** oilseed rape, mustard, vegetables (cabbage, cauliflower), papaw **Animal feed** fodder
Sapindales	**Food** orange, lemon, lime, mango, cashew, pistachio, lychee, maple sugar

Source: based upon Henry (2005a)

better management of plant production (e.g. improved plant nutrition using fertilizers). The development of human societies as we know them today began with the domestication of plants (and animals) more than 10,000 years ago. The expansion of human societies has relied largely on the continued genetic selection of these species to refine them to satisfy our demands for more food and better quality food. Evolving approaches to the management of agriculture that have advanced the whole crop or animal production system has complemented genetic strategies to deliver these outcomes.

Estimated global production of foods for 2004 (FAO, 2007) is as follows:

Cereals	2270 million tonnes
Fruits & Vegetables	1384 million tonnes
Roots & Tubers	718 million tonnes
Pulses	61 million tonnes
Oilseeds	142 million tonnes
Sugar Crops	1577 million tonnes
Milk	622 million tonnes
Meat	260 million tonnes
Eggs	3 million tonnes

The parts of plants that we eat vary widely. The seeds are generally highly nutritious because they represent a concentrated source of energy and nutrients for a new plant itself. In many cases this has been enhanced by human selection and domestication. Fruits (the part of a plant surrounding the seed) have evolved to be attractive to animals to aid plant seed dispersal and we have also further developed this feature by human selection. Roots and tubers and sometimes even stems are plant organs for the storage of energy, often making them attractive foods. The leaves and flowers of many species are also eaten but are generally less nutritious or attractive as foods. Major plant food groups contribute to human food needs in specific ways. For example, the cereals are the primary sources of carbohydrates or energy in most human diets, while grain legumes or pulses provide significant amounts of protein.

Cereals

Cereals are the major source of calories or energy in many human diets.

The three leading cereal species – wheat, rice and maize – are each produced in large quantities (600–700 million tonnes of each per year) and account for a large part of all human food measured in calories (energy) or protein. More maize is produced, but much of it is feed to animals and only contributes to human diets indirectly. In contrast, most wheat and almost all rice are consumed directly by humans. The wheat is milled to produce flour which is used in a wide range of foods such as bread and pasta, while rice is largely consumed as a whole grain food. Barley and sorghum are fourth and fifth ranked in production (Figure 2.1).

Box 2.1 *Cereals*

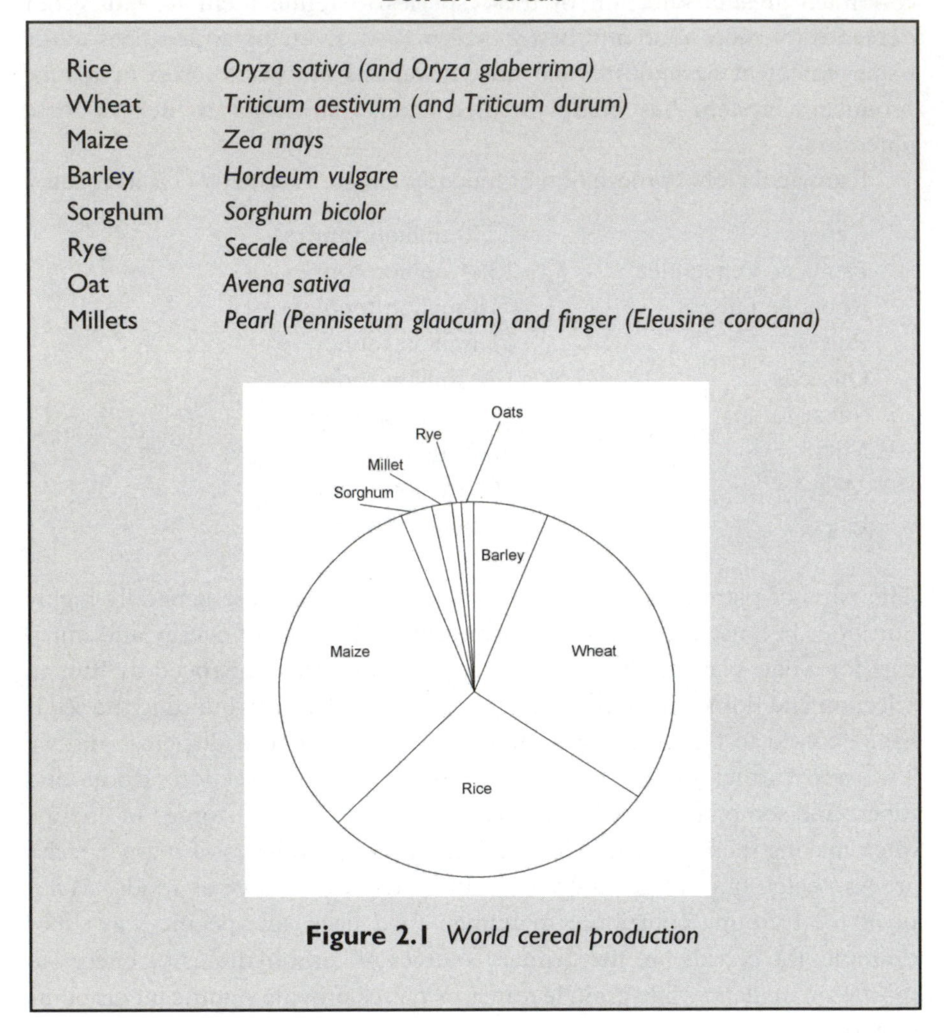

Rice	*Oryza sativa (and Oryza glaberrima)*
Wheat	*Triticum aestivum (and Triticum durum)*
Maize	*Zea mays*
Barley	*Hordeum vulgare*
Sorghum	*Sorghum bicolor*
Rye	*Secale cereale*
Oat	*Avena sativa*
Millets	*Pearl (Pennisetum glaucum) and finger (Eleusine corocana)*

Figure 2.1 *World cereal production*

Daily average global calorie intake from rice is 557, wheat 521 and maize 117, representing more than 40 per cent of the total of around 2800 from all foods (FAO, 2007). Rice has been the traditional staple food in Asia and maize in the Americas, with wheat widely consumed elsewhere. Wheat was domesticated in the Fertile Crescent and rice in Asia (China). Globalization of food consumption is resulting in all three species being consumed as a significant part of diets worldwide. The per capita consumption of cereals is likely to decline in some regions as diets become more international and change to include more meat and other foods. Migration has spread dietary habits globally, and combined with high levels of international travel, contributed to consumption of many foods on a widespread global basis. Cereals have been adapted to the production of a very wide range of human foods (Henry and Kettlewell, 1996). For example, wheat is used to produce the following types of products:

- fermented (leavened) breads (including pan bread, sandwich buns, hearth breads, sweet goods, Danish, croissant and raisin bread, and steamed breads);
- flat breads and crackers (including chapatti, roti, naan, paratha, poori, balady, pita, barabari, tortillas, pizza crust, English muffin, crumpets, bagels and pretzels);
- cookies and cakes;
- noodles (including instant noodles, white salted noodles, yellow alkaline noodles and Udon noodles);
- pasta;
- breakfast foods;
- starch and gluten as food ingredients.

The production of such a diversity of foods from a single plant species emphasizes the narrow genetic base of most human food and our great dependence on cereals for food.

Cereals account for the bulk of energy or calories in many human diets and are as a consequence important sources of total protein. However, the proteins in cereals are of relatively poor nutritional quality, requiring other foods to provide balance in the diet. In addition to the seeds of the grasses (cereals), our primary source of carbohydrate, we eat the seeds of legumes (for protein) and those rich in oils (oilseeds).

Pulses

Grain legumes or pulses are protein rich seeds consumed as a complement to cereals, especially in many largely vegetarian diets such as those commonly consumed in countries such as India. Increased consumption of pulses rather than meat as a protein source is an important option to consider in efforts to satisfy food needs from a smaller agricultural footprint. These crops are critical for food production in other ways; through associated micro-organisms they fix and contribute nitrogen to the soil, enhancing the yields of other crops such as cereals grown in rotation or following the legume crop. The contribution of these crops to sustainable food production is often seriously underestimated because this factor of nitrogen fixation (as a substitute for fertilizer use) is not considered.

Box 2.2 *Pulses*

Pulses include: Adzuki beans, Alfalfa, Beans (Green), Black gram, Borlotti bean, Broad beans (Faba beans), Carob, Chickpea, Haricot beans, Kidney beans, Lentil, Lupins, Mung Bean, Pea, Peanut, Pigeon pea, Pinto bean, Soybeans, Tamarind, Vetches. These foods are the seeds of plants in the legume family (Fabaceae). They represent an important source of protein in human diets. Many of these species contain toxic or anti-nutritional factors that have been reduced by breeding selection, or are reduced by processing or cooking. For example, pulses may contain trypsin inhibitors (compounds that inhibit the action of trypsin, a protease (protein degrading enzyme) produced in the digestive system, and as a result reduce the ability of the human or animal to gain nutrition from protein in the diet).

Oilseeds

Oilseeds provide an important contribution to human diets. Canola, a product of modern plant breeding, has recently emerged as a major oilseed. Developed in the early 1970s, canola was produced by selecting cultivars of rapeseed that were low in both erucic acid and glucosinolates to reduce problems of rancidity and potential toxicity of the oil. Sunflower domesticated in North America has also been an important oilseed. Oil palms have been a major source of food oil from more tropical regions. Soybeans are a significant oil source especially in North America. Recent genetic improvement of soybean has targeted the mutation of three genes to reduce greatly the linolenic acid (18:3) content to avoid the formation of undesirable trans fats during the stabilization of the wild type soybean oil by hydrogenation. This is a good example of how humans have modified oilseeds to suit our food needs. We now aim to combine the desired function of the oil in the food in terms of physical properties with desirable nutrition and health attributes.

Fruit and vegetables

Plant parts other than seeds are also major foods. Fruit and vegetables are at least minor parts of most human diets. However, some fruits are very important regionally. For example, the banana, originally domesticated in South East Asia, probably having a centre of diversity in New Guinea, is a staple food in Uganda and is important in other countries. Fruits such as grapes and citrus are consumed widely throughout the world. Other fruits, especially many tropical fruits, are consumed on a much more regional basis.

Roots and tubers

The roots and tubers of plants are also important components of many human diets.

The potato is widely consumed as a staple food; originally from South America, it has become very important elsewhere. Infamously, the outbreak of disease in potatoes in Ireland led to major food shortages in the mid-1800s. Potatoes have been ranked third in importance as a food, after wheat and rice. Increased emphasis on potato production contributes to food security by helping to diversify food production and reduce reliance on cereals. The potato can be produced in many regions and has been more stable in price because it is produced locally and has not become an exported commodity. Use of potatoes to substitute for some of the wheat in bread is an option for increased use of potato in human diets.

Cassava, originally from Brazil, is a starchy tuber crop that is especially important as a food in Africa. Yams (*Dioscorea* species) are also important root crops in many tropical regions.

Sugar

The attractiveness of sweet foods has made the cultivation of crops as a source of sugar an important activity. Sugarcane in tropical areas and sugar beet in temperate regions are import food crops. Sugarcane was domesticated in South Asia and New Guinea, and globally it makes a significant contribution to the energy in human diets (averaging 202 calories per day).

Beverages from plants

Humans consume several beverages made from plants. Tea and coffee are widely consumed, while alcoholic beverages are produced from several plants. Barley and to a much lesser extent sorghum and wheat are used to produce beer by fermentation of the carbohydrates. Grapes and rice are used to produce wines (sake in the case of rice). Fermented plants are distilled to produce alcoholic beverages such as whisky. Alcohol is a significant but not necessarily desirable part of human food energy consumption.

Food from animals

Animals were domesticated by humans at the same time as plants (Diamond, 2005a). Cows, pigs, sheep and goats were early domesticates and continue to be important sources of food. The potential for the domestication of a range of both plant and animal species in the one region probably explains the location of early domestication and the beginnings of agricultural communities in the Fertile Crescent, and later separately in several other key regions globally.

Food from animals also depends upon plants. Animals graze plants in pastures or are fed grain in more modern intensive production. Eggs are produced from chickens fed on grain in similar production systems. Fish production from the oceans has declined rapidly due to overharvesting. Increasingly, fish is produced by farming or aquaculture, using increasing amounts of plant-derived feeds. However, the human consumption of food from animals is a relatively inefficient mechanism of obtaining nutrients from plants. Food energy sourced from animals requires relatively large amounts of plant material as animal feed (around ten times) relative to the amount of plant material that would be required to supply the energy in direct plant consumption by humans. Increasing global consumption of meat associated with increased affluence is resulting in a greater per capita environmental footprint. Animals also contribute methane, a significant greenhouse gas, to the atmosphere. A higher proportion of animal food in human diets results in a greater per capita impact on the environment due to increased methane production. In this way the choices of food consumers have a large impact on the global environment.

Maize is the grain crop produced in greatest quantity globally, but most maize is fed to animals. The amount of maize produced currently exceeds the amount of wheat or rice, the two major food crops consumed directly by humans.

Human food preferences and plant domestication

Domesticated plants have been selected for genetic attributes that suit human purposes. The evolution of human populations has necessarily been influenced by food selection and availability. We have selected plants that we find pleasing (taste good) and that can be produced conveniently (easy to grow and prepare to eat). This explains the forces that have shaped the selection of our domesticated crops, but how might plants have influenced human evolution? Humans that prefer or select food that has a high nutritional value and contributes to good health are more likely to survive and pass on their genes than those that favour foods that do not meet physiological and nutritional requirements. In this way domesticated plants and humans co-evolved to be dependent upon one another. Humans seem well adapted to the omnivorous diet of our hunter-gatherer ancestors. We may not be so well suited to our modern domesticated plants that have been selected for human tastes which

Box 2.3 *Barley: the first plant domesticated*

Agriculture and the growth of modern human societies began more than 10,000 years ago. Barley has been widely considered as the plant likely to have been domesticated first. Wild barley (*Hordeum spontaneum*) is still abundant in the Fertile Crescent (Figure 2.2) and can be seen growing on the sides of the citadel (inset) in Aleppo, Syria. Wild and cultivated barley continue to interbreed (Bundock and Henry, 2004). Domestication of barley and the other cereals was based upon selection of this large seeded grass that grew in large populations that could be easily harvested. Wild barley shatters; that is, the seed falls from the plant when ripe to ensure dispersal and survival of the species. However, domestication has involved selection of barley that does not shatter but retains its seed, so that it can be harvested at one time by humans. Selection for similar genetic changes

Figure 2.2 *Wild barley. Wild barley plants are shown growing in a field near Aleppo, Syria*

has been associated with the domestication of other seed crops such as rice, wheat and maize. Aleppo, together with Damascus, are claimed to be the longest continuously inhabited existing cities probably owing their development to the societies established following the foundations of agriculture using the wild barley. The barley plant may have been the first species domesticated in the Fertile Crescent more than 12,000 years ago.

may have largely been developed for survival in a pre-agricultural environment. We now have human societies that are overwhelmingly dependent upon domesticated plants for survival. Current world populations greatly exceed the numbers that could survive on the planet, without agriculture, as hunters and gatherers of food. Furthermore, we now exceed the populations that could survive by cultivation of wild plants and depend upon the characteristics of our domesticated plants for survival. It is worth examining the process of domestication and the ongoing development (more extreme domestication of these plants) and the potential for additional domestication of species that have not yet been domesticated by humans. Chapter 10 will consider these issues in more detail.

Much of the selection during plant domestication and breeding has been for characteristics which allow the production and harvest of large quantities of the edible part of the plant. Maize was domesticated in Meso-America more than 7000 years ago (Thompson, 2006). The wild teosinte plant from which maize was domesticated is very different in appearance (Table 2.2). Domestication involved changing the architecture of the plants. Ease of production has been an important consideration driving human domestication of plants.

Plants have been selected by humans on the basis of their attractiveness as foods – taste and texture have probably been the key influences on human behaviour in selection of foods. The ready availability of food that tastes good has clearly been an important incentive for humans in the development of agriculture.

Table 2.2 *Domestication of maize*

Primitive maize	Domesticated maize	Common traits
Slender cobs	Loss of dormancy	Unisexual inflorescences
Short ears	Increased grain size	Tassle and ear
Hard grains	Loss of hard casing	C4 photosynthesis
Brownish colour		

Source: Henry, 2001a

Box 2.4 *Human selection of rice as an attractive food*

Domesticated rices have many attributes that have been selected by humans because of their attractiveness in a food (Bradbury et al, 2008). These include texture, aroma and appearance.

Aroma or fragrance (Bradbury et al, 2005a) Recent research has identified that a loss of function of a gene in rice leads to the highly desirable aroma or fragrance of Thai and Basmati rices. The fragrant rices of Thailand are very different to the Basmati rices of India and Pakistan, but they both share a common fragrance gene suggesting a common ancestor contributed the fragrance to many of these divergent rices. Fragrance is due to a compound, 2-acetyl-1-pyrroline, that accumulates in parts per billion in the plants as a result of the mutation selected by humans. Apparently humans have selected this trait because they have an ability to detect the smell of this compound at very low concentrations and find it highly desirable. Modern plant breeders have progressed slowly with the development of fragrant cultivars, until the recent advent of DNA-based tests, probably because of the genetically recessive nature of the trait (requiring the gene to be carried by both parents for it to be expressed in the rice plant) and the difficulty of detecting the fragrance of individual seeds or plants. Most rices result from a single selection event, but recently other mutations in this gene have also been found to result in loss of function and the associated fragrance, indicating several separate human selection events.

Recent research has revealed a remarkable twist to this story. Fragrant rices have been found to be highly susceptible to salt stress. Humans are cultivating a plant that is very poorly adapted to survival under environmental stresses because of a defect in metabolism that also happens to give the plant a highly desirable taste. This provides an excellent example of the conflict between traits desirable to humans and favoured under domestication and traits that favour survival in wild populations. The detection of a single fragrant grain is difficult for most human noses and even modern analytical chemistry. This indicates just how desirable this trait must have been to those originally selecting it. Fragrance has been difficult to assess on individual grains of rice in breeding, but can now be assessed by DNA analysis (Bradbury et al, 2005b).

Cooking requirements (Waters et al, 2006) Two different mutations have been found in some domesticated rices, each of which allows the rice to be cooked at a temperature around 8°C lower than wild rice. Human selection for these mutations has allowed rice to be cooked at lower temperatures and to produce a rice grain with more desirable texture. This selection happened at least twice, possibly at different times or locations, to account for these two different genetic forms, illustrating the desirable nature of this trait.

The domestication of many plants remains a mystery with limited evidence as to the identity of the wild plants from which they were domesticated or the time of location of domestication. Some major crop plants for which we do have some of this type of information have origins that involved complex or poorly understood genetic processes. For example, common bread wheat is hexaploid (six copies of each chromosome as compared to a diploid like a human with two sets of chromosomes – one from each parent (mother and father)) that has resulted from the combination of three different grass species (Figure 2.3). Cultivated sugarcane has a very large number of chromosomes (greater than 100) and is also the progeny of crosses between different wild species (Dillon et al, 2007). Many cultivated plants have multiple copies of their genes and chromosomes that have arisen by processes other than the crossing of different parental species. The presence of multiple copies of related chromosomes (polyploidy) is a feature of many flowering plants and is probably especially common in cultivated plants. Polyploidy may be a significant advantage to the plant and is often associated with enhanced plant

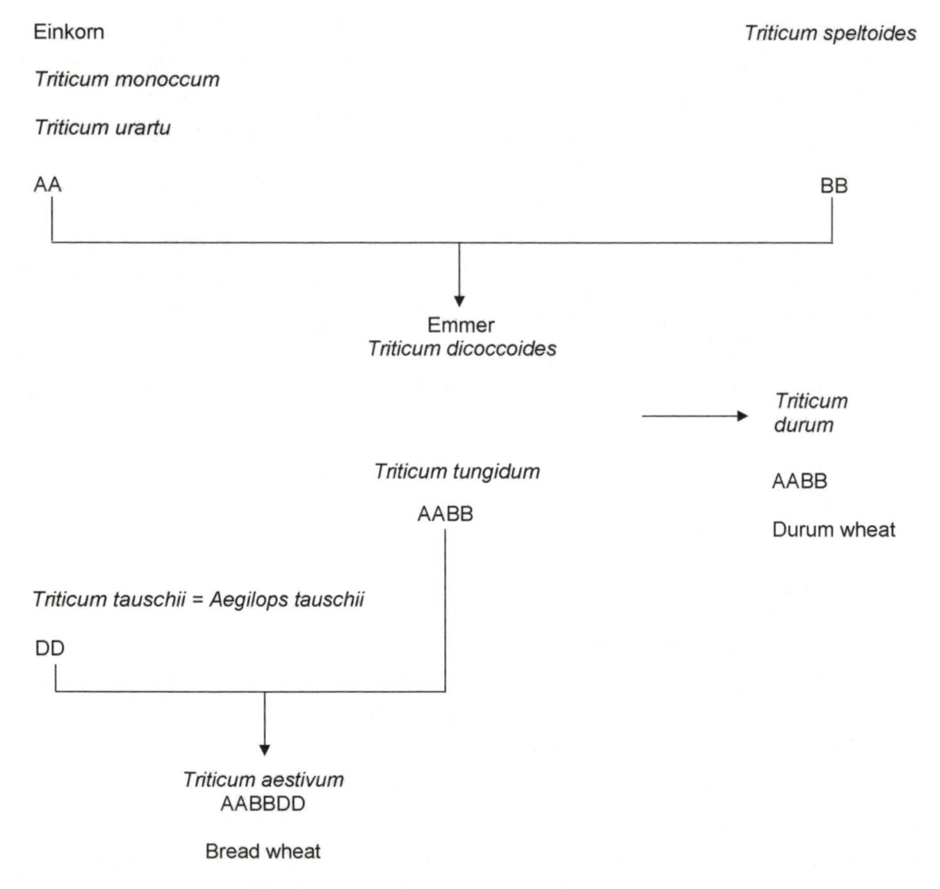

Figure 2.3 *Genetic origin of bread wheat*

performance and adaptation to environmental variation. Domestication of the peanut involved human cultivation of a polyploidy plant distinct from the diploid plants in wild populations. The potential for further domestication of plant species to meet human needs for food, feed, fibre and energy in the future is considered in more detail in Chapter 10.

How much do we eat? (growth in food consumption)

The growth in food consumption is being driven by several key factors:

1. Growth in human population is a primary cause of continued growth in demand for food.
2. However, economic growth in countries such as China and India is also driving increases in per capita consumption of food. Increased travel and communication is also changing food preferences, with diets becoming more global and many foods being marketed internationally.
3. Increased consumption of animal products is driving strong growth in demand for feed. China is a country with a very large population undergoing a rapid rise in incomes, and a dramatic increase in meat consumption has resulted. Consumption of fish, pork, poultry and beef has more than tripled on a per capita basis since 1970. Human and animal feed are competing uses for crops. It takes more than 4kg of grain to make a kilogram of poultry and around 20kg to make a kilogram of beef. Meat consumption globally is expected to double by 2050 (Roberts, 2008). Changes in food preferences may see a reduction in rice consumption and a strong growth in the consumption of meat, eggs and fish in Asia (Von Braun, 2007).

A major ongoing issue is the extent of geographical mismatches between supply and demand. Food transport consumes significant energy. Efforts to better match food production to local demand have potential to make a significant contribution towards satisfying the growing demands on agriculture. The concept of food miles has emerged as a measure of the distance food has been transported and promoted as a way consumers can support the consumption of local produce with less environmental impact due to the energy costs of long-distance transport. This concept is only valid if the food is produced efficiently. Two examples of the problem are the risk of using large amounts of energy to grow tropical plants in a cold climate closer to markets, or the displacement of biodiversity to produce crops at lower yield in a less favourable environment closer to the consumer. We need to be able to define food labelling options that also allow for these factors and provide

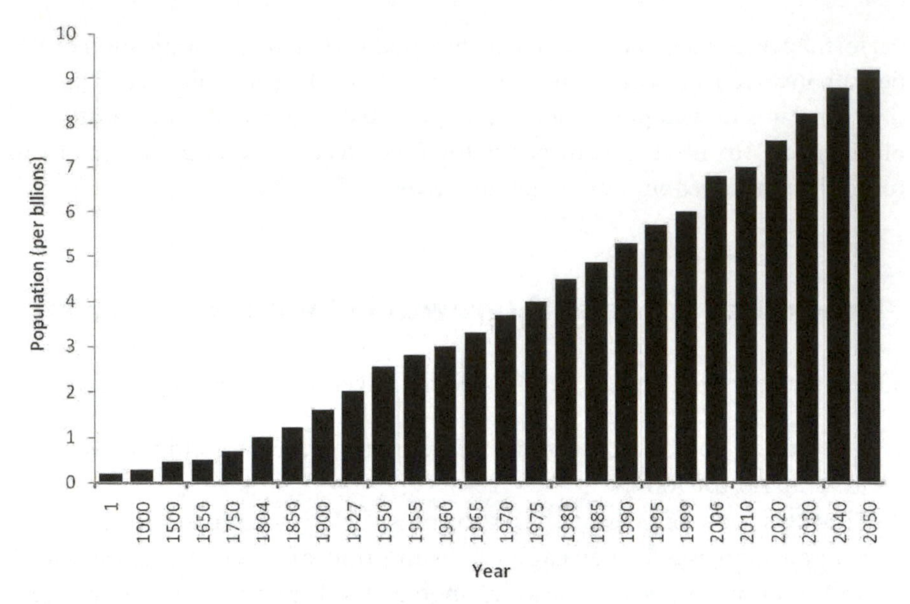

Figure 2.4 *World population growth*

reliable guidance to consumers in choosing food that will reduce their environmental footprint. A measure of the total energy consumed, or better the total greenhouse gas generated in their production, is a measure that incorporates many of these environmental considerations.

The growth in human population (Rosenberg, 2008) is depicted in Figure 2.4. The rate of growth of human population is beginning to slow. Ongoing population growth is driven partially by the age structure of the population. A stable population requires that each adult has an average of slightly more than one offspring to allow for the impact of premature deaths (e.g. accidental deaths before reaching reproductive age). Many developing countries with a large proportion of young people will experience continued strong population growth even if family sizes are restricted as the young reach reproductive age. Developed countries such as many in Europe are expecting significant population declines as the population ages.

The consumption of food per capita is also a major cause of growth in food consumption globally. The contribution of growth in numbers and growth in consumption per capita varies greatly in different parts of the world. However, these two factors are probably of about equal importance currently at the global level. While population growth continues, its rate of growth is slowing as societies become more affluent. Economic development is also associated with major changes in dietary habits (WHO, 2008). The recent rapid growth in per capita food consumption in many countries has

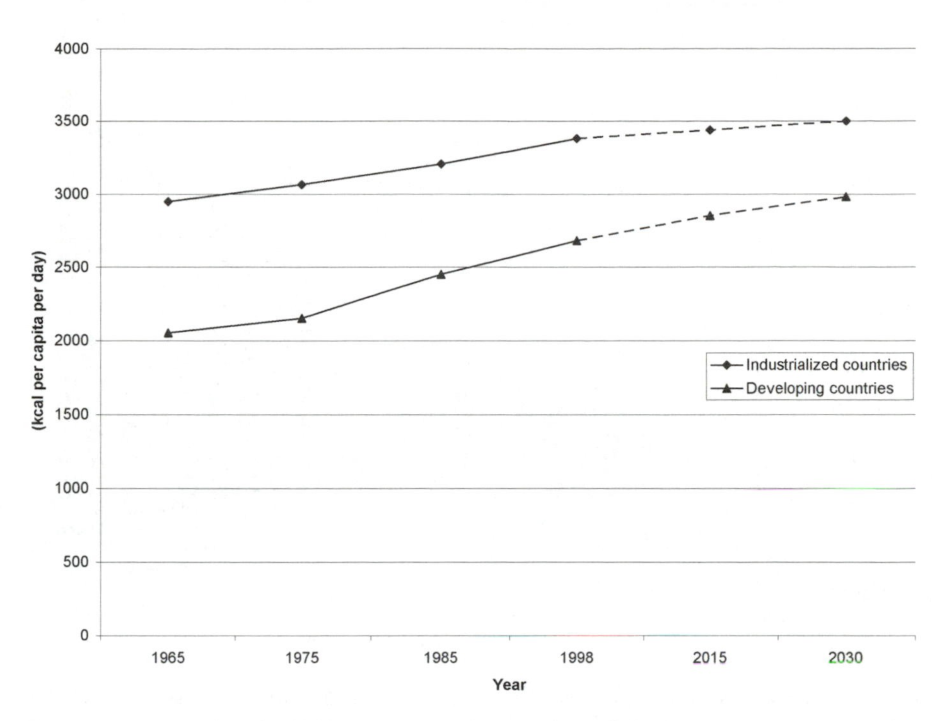

Food consumption (WHO, 2008, www.who.int/nutrition/topics/3-foodconsumption – accessed 7 October 2008)

Figure 2.5 *Global food consumption by industrialized and developing countries*

seen them begin to catch up with consumption patterns in the developed world (see Figures 2.5, 2.6 and 2.7).

This phase of increased consumption may slow as more of the world population reaches high levels of food intake. These changes in dietary habits are not all positive and many people in the world now consume too much food. The inequalities in food supply globally result in large numbers continuing to face food deficits, while probably equally large numbers consume excess food. Obesity is a major concern in developed countries, with a wide range of diseases increasing in frequency as food intakes become excessive. It has been suggested that life expectancies may begin to decline if this problem is not addressed. The number of people that are obese may exceed 2 billion by 2012 (Biospectrum Asia, 2008). This has resulted in considerable private efforts to produce drugs that will treat this problem. Greater public efforts to promote healthy dietary habits are needed. This suggests that we need to focus on the production of plants that will support healthy eating habits and are attractive to consumers if we want to change food consumption patterns to support healthy human populations.

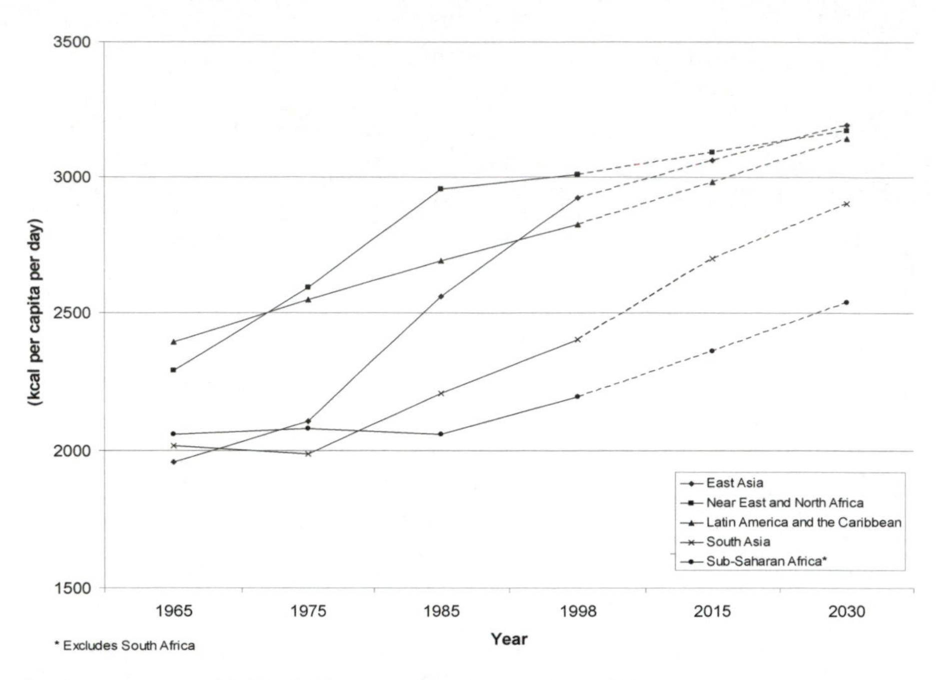

Food consumption (WHO, 2008, www.who.int/nutrition/topics/3-foodconsumption – accessed 7 October 2008)

Figure 2.6 *Global food consumption by region*

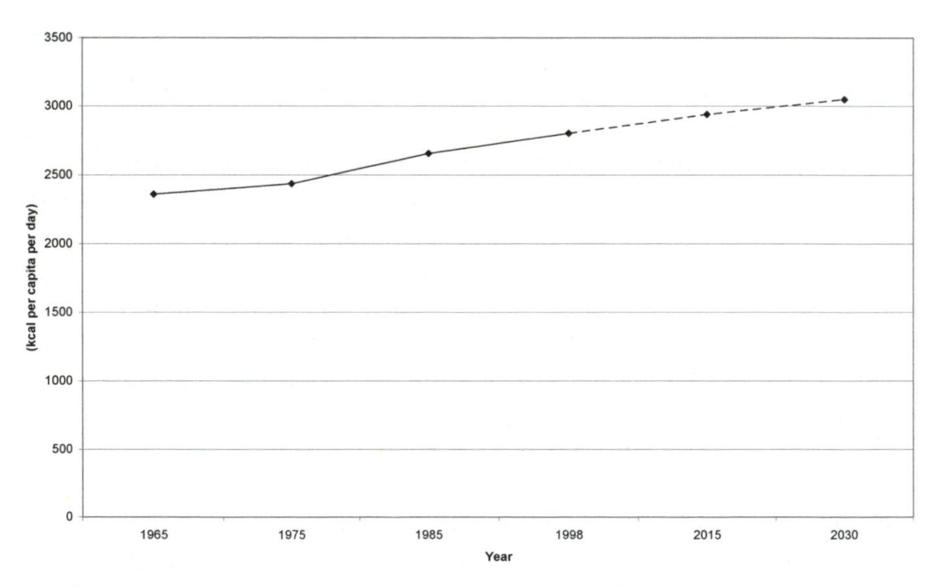

Food consumption (WHO, 2008, www.who.int/nutrition/topics/3-foodconsumption – accessed 7 October 2008)

Figure 2.7 *Global food consumption – world*

Food versus non-food uses of plants

Human demand for plants for non-food uses also continues to grow – plants are used for a wide range of non-food uses. The use of plants for fibre and construction does not usually compete with food uses as woody plants or the woody parts of plants are not usually edible. Some uses of plants for food and fuel may compete and these are the focus of considerable attention because of the conflicts they may create. For example, the use of the starch from maize, potatoes or cassava for fuel production may compete directly with human food use and maize, and sorghum use may compete with animal feed uses that indirectly contribute to human food. This will be discussed in more detail in Chapter 6. Grain production to feed animals does more often compete with grain production for direct use as human food. Competition in the use of land or water is possible for many uses of plants.

A major use occupying a substantial area of land is production of wood products in forests. Global forest production estimates for 2004 are as follows (FAO, 2007):

Round Wood	1646 million m^3
Fuel Wood	1772 million m^3
Charcoal Wood	44 million tonnes
Sawn Wood	416 million m^3
Paper (including cardboard)	354 million tonnes

Forest products supply energy (fire wood and charcoal) for cooking and heating, timber for construction, and fibres for paper and cardboard.

Electronic communications may reduce demand for newsprint but the use of printer paper remains widespread. Construction from wood may be favoured as a mechanism of carbon storage or capture, especially when compared with the carbon emissions associated with alternative construction materials.

Composition of plants for food and non-food uses

Plants are composed of water, some inorganic salts and a wide range of organic molecules, ranging from very simple compounds to large macromolecules. Plants also contain essential vitamins.

Carbohydrates are the major component of plants and are a key nutrient in staple foods such as cereals, providing the bulk of the calories in human diets. These carbohydrates include simple sugars and more complex carbohydrates often in the form of starch (a polymer of glucose). The most common sugar in plants is the disaccharide sucrose (the sugar of sugarcane or sugar beet) used widely as a sweetener in human foods. Humans and animals are well equipped to use sugars and starch as a source of energy. Human saliva contains the enzyme alpha-amylase that begins the digestion of starch in the mouth. This demonstrates the way in which humans and animals are adapted to consume foods rich in plant starches.

Proteins are also an important part of plants from a food value perspective. Plants synthesize amino acids (constituents of proteins) that are not able to be produced by the metabolism of humans or animals. These 'essential' amino acids must be obtained in the diet. A balanced human diet needs to include plant or animal components that provide the amino acids needed to produce proteins essential to normal cellular function. Animals with proteins closer in amino acid composition to human proteins are often a better balanced source of amino acids for human diets than plants.

Fats sourced from plants are generally desirable in human diets. A range of crops are grown for their oil content: soybean, canola, sunflower, peanut, castor, olive, safflower, coconut and oil palm.

A unique feature of plants is the presence of a cell wall that provides a rigid structure for the plant. Plant cell walls contain lignin and a range of polysaccharides (carbohydrates). Cellulose is a cell wall polysaccharide that forms microfibrils that are an important structural component of the cell wall. These cellulose fibres are surrounded by non-cellulosic polysaccharides of differing composition in different plant species. Some of these cell wall polysaccharides have an important role in human diets as dietary fibre and are food for bacteria in the digestive systems of animals. This is most developed in the ruminants such as cows that are able survive on a diet high in these structural carbohydrates. Even humans have large numbers of bacteria in the large intestine and bowel that are able to partially digest these polymers. An average healthy human adult probably has more than 1kg of bacteria in their gut active in the digestion of food. Plants with a high sugar, starch or protein content are likely to be good food plants, while plants or parts of plants with a high cell wall or lignin content, as in the woody parts of plants, are generally not edible but may be useful for other purposes such as timber or paper. The targeting of plants to uses based upon their composition is an important consideration when competition for end use becomes an issue. The composition of plants in relation to their suitability for energy use will be described in more detail when we consider energy uses of plants in Chapter 5.

Genetic resources for food crops (conservation and utilization)

The genetic resources available to support the sustainable production of food and for other traditional uses are the primary, secondary and tertiary

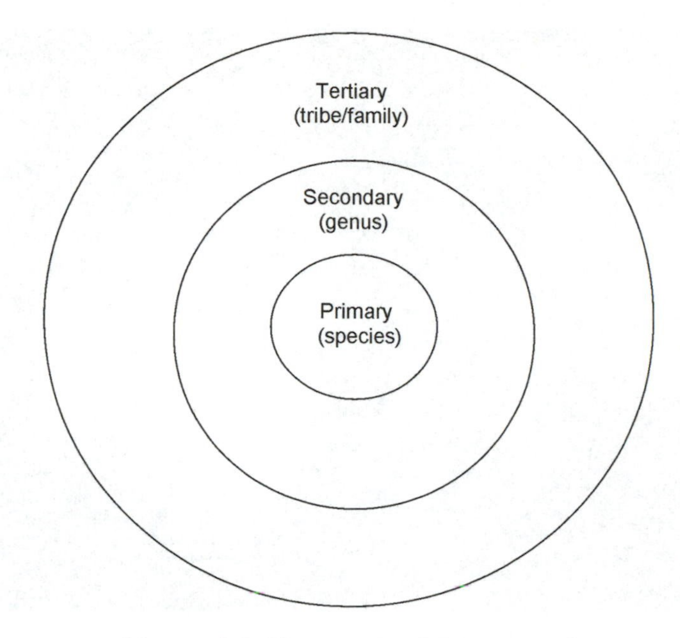

Figure 2.8 *The concept of the gene pool*

gene pools of the key species (Figure 2.8). The primary gene pool includes those individuals with which the plant can freely inter-breed. The secondary gene pool includes plants that are less closely related and would not normally inter-breed in nature, but that can be used in plant breeding. The tertiary gene pool extends to species that may often only be used in plant breeding by using advanced or novel breeding strategies and technologies, including gene transfer (genetic modification) techniques. We may consider cultivated plants to have both a domesticated and a wild gene pool. The wild gene pools of crop species are often referred to as wild crop relatives. These plants in the wider gene pool provide a reservoir of genetic variation that may be called upon to ensure continued production of essential crops in the face of threats from new diseases or environmental change.

Wild crop relatives are an important group of plants that deserve greater efforts directed at *in situ* (where they are in wild populations) conservation and *ex situ* (in seed banks, living collections and DNA banks). Wild populations of crop species are threatened by all the factors that threaten other species, but may also be subject to the risk of being genetically impacted by gene flow from domesticated crops. The cultivation of large genetically uniform domesticated crops close to small populations of wild individuals of the same species risks the wild populations being genetically overwhelmed by pollen flow from the domesticated plants, with the resulting potential loss of wild genes. This is a possible problem with Macadamia in Australia, with

Figure 2.9 *Wild grape and rice relatives*

large clonal plantations being grown within the range of the native species which are rare in the wild. A Macadamia conservation trust has recently been formed to support conservation of the wild genetic resources of these species. Examples of wild relatives of crop species are shown in Figure 2.9.

These members of the tertiary gene pool of grape and rice provide knowledge and possibly genes that might allow these crop species to be adapted to environments outside the current range of the crop species.

A distant relative of the grape, *Cissus antarctica*, is found in the sub-tropical rainforests of Australia (upper panel). DNA analysis shows that this species and three other species from the same regions are much closer to the grape genus *Vitis* than other *Cissus*. Cultivated grapes are very susceptible to fungal diseases when grown in these warm and wet environments. A wild rice relative, *Potamophila parviflora*, grows in the rivers of central Eastern Australia, extending inland to grow in cool winter temperatures in the upper reaches of these river systems (lower panel). DNA evidence indicates that this species' closest relative is the North American wild rice (*Zizania* species).

Two examples of the importance of wild crop relatives are described below – rice and sorghum. Rice is a major food crop and the security of its supply is very important for many human populations. Wild crop relatives have a central role in providing a source of genetic variation that can be used to maintain production in response to biotic (e.g. new diseases) or abiotic (e.g. climate change) challenges. Sorghum is also an important human food crop, but in its current form is not as highly valued as rice. The wild relatives of sorghum offer not only a resource for ensuring security of supply, but also a source of new variation that might be used to make sorghum more attractive as a human food.

Wild relatives of rice and potential for use in rice improvement

Rice is one of 22 species in the *Oryza* genus (Table 2.3). This genus is one of 12 in the Oryzeae tribe (Table 2.4). *Oryza sativa* and other closely related species that can be readily cross-pollinated with rice (designated A genome species) represent the primary gene pool of rice. Genes in other species in the genus may be accessed for rice improvement and represent a secondary gene pool. The wider genus may be considered a tertiary gene pool for rice from which genes could only be accessed with much greater difficulty. Use of genes from these wild relatives of rice requires an understanding of the relationships between the rice species (Kovach and McCouch, 2008). The wider gene pool of rice as represented in the wild relatives is an important target for conservation of biodiversity to support global food security. This gene pool remains relatively poorly characterized and under-utilized to date. This would also be the case for most important food crop species. The *Oryza sativa* complex includes those closely-related species that can be directly inter-bred with rice. These and other species in the genus are currently being characterized by

Table 2.3 *Rice genus*

Oryza sativa complex
O. sativa L.
O. nivara Sharma et Shastry
O. rufipogon Griff.
O. glaberrima Steud
O. barthii A. Chev.
O. longistaminata Chev. Et Roehr.
O. mederionalis Ng.
Oryza officinalis complex
O. officinalis Wall ex Watt
O. minuta Pesl. Et Presl.
O. rhizomatis Vaughan
O. eichingeri Peter
O. punctata Kotchy ex Steud
O. latifolia Desv.
O. alta Swallen
O. grandiglumis (Doell) Prod.
O. australiensis Domin
O. brachyantha Chev. Et Roehr.
O. schlechteri Pigler
O. ridleyi Hook. F.
O. longiglumis Janse
Oryza meyeriana complex
O. meyeriana (Zoll. Et Mor. Ex Steud.) Baill.

genome sequencing and analysis, with the closest relatives being given the greatest attention. The more diverse species within the tribe provide options for more radical and probably longer-term genetic improvement of the cultivated plant species.

Table 2.4 *Rice tribe*

Genus	Species	Distribution
Oryza	22	Pantropical
Leersia	17	Worldwide
Chikusichloa	3	China, Japan
Hygroryza	1	Asia
Porteresia	1	South Asia
Zizania	3	Europe, Asia, North America
Luziola	11	North and South America
Zizaniopsis	5	North and South America
Rhynchoryza	1	South America
Maltebrunia	5	Tropical and Southern Africa
Prosphytochloa	1	Southern Africa
Potamophila	1	Australia

Wild relatives of sorghum and their potential for use in crop improvement

Wild relatives of some crop species represent a very valuable resource that can be used to expand the gene pool of the domesticated plant. Cultivated sorghum originated in Africa and was probably domesticated 5000–6000 years ago. The sorghum genus includes 25 species and those species that have not been domesticated have many characteristics that would be of value in a crop plant. They are tolerant of extreme heat and drought and grow with limited nutrients. Hybrids between cultivated sorghum and some of these wild species have been recently produced. The availability of a completed sequence of the sorghum genome (Paterson et al, 2009) provides access to tools that would allow molecular-assisted breeding (the use of DNA analysis to support plant selection). Sorghum genetic improvement is just beginning to benefit from these new tools and insights into genetic relationships between domesticated sorghum and wild relatives (Dillon et al, 2007). Wild species that have been shown to be the closest relatives of cultivated sorghum have been successfully cross-pollinated with sorghum, confirming the DNA evidence of a close relationship. This is a good example of how DNA analysis of plant relationships can guide conventional breeding to successfully explore more options for accessing novel genetic variation.

Genetic resources for crops species and their wild relatives need to be conserved in the wild (*in situ*) when possible. However, this is often not possible and seedbanks of the major species hold significant numbers of cultivars and other genotypes (Table 2.5) in *ex situ* (not in the wild) collections. Seed life can usually be extended by keeping the seeds dry and cool. Cool storage using refrigeration depends upon a constant energy supply that may be disrupted by major disasters. Long-term conservation of large collections of seeds is being ensured by storage in the arctic permafrost. The seeds of many species (e.g. many from rainforests) do not survive drying and storage, and they must be conserved *ex situ* as living specimens. *Ex situ* plant conservation options for plants in general (not just relatives of cultivated food crops) are discussed in more detail in Chapter 7. For cultivated plants, farmers fields remain a very important location for the conservation of diversity as farmers' continue to grow traditional varieties and retain the seed from previous crops. These crops will often retain diversity that may be lost in the more intense genetic selection of elite varieties in plant breeding.

DNA banks are a new option to support the more efficient use of the living collection or seedbank. DNA in these collections can be analysed to identify genotypes that contain useful genes or genetic characteristics. These

Table 2.5 *Seedbanks for major crop germplasm resources*

Species	Number of accession in collections
Wheat	784,500
Barley	485,000
Rice	420,500
Maize	277,000
Oat	222,500
Soybean	174,500
Sorghum	168,500
Apple	97,500
Millet	90,500
Cowpea	85,500
Peanut	81,000
Tomato	78,000
Chickpea	78,000
Pea	72,000
Capsicum	53,500
Grape	47,000
Triticale	40,000
Sweet potato	32,000
Potato	31,000
Faba bean	29,500
Lupin	28,500
Cassava	28,000
Rye	27,000
Lentil	26,000
Onion	25,500
Pigeon pea	25,000
Yam	11,500
Carrot	6000

Source: Henry, 2005a

(www.fao.org/ag/AGP/AGPS/Pgrfa/swrshr_e.pdf accessed 16 February 2009)

can then be sourced from the seedbanks and used in agricultural production or in plant breeding. International seedbanks make seed available to plant breeders worldwide for use in developing new varieties.

Technologies and strategies for the improvement of agricultural species

Humans have continuous genetically 'improved' domesticated crop plants by processes of both conscious and unconscious selection and breeding. This genetic change has targeted the growth characteristics of the plant (higher yields), and especially the attractiveness of the food to human taste and the convenience of harvest and processing. Selection by screening of large

Box 2.5 *DNA banks for conservation and support of plant improvement*

DNA banks (De Vicente and Andersson, 2006) represent a new type of *ex situ* conservation that does not attempt to provide for continued propagation of the plants, but stores DNA for analysis in plant identification, genetic diversity studies, population genetics and evolutionary analysis. The DNA banks allow research aimed at gene discovery in support of plant improvement. Such banks that mirror seed banks or other living collections can be used as primary tools for screening and selection of germplasm in plant breeding applications. The Australian Plant DNA Bank includes DNA of important food species such as wheat, barley and rice that are derived from seed samples held in international seed banks.

DNA Bank	Website
Missouri	www.mobot.org/MOBOT/reserach/diversity/dna_banking.htm
Kew	www.kew.org/data/dnaBank/homepage.html
South Africa	www.nbi.ac.za/
Brazil	www.jbrj.gov.br/pesquisa/div_molecular/bancoda/index_ing.htm
Australia	www.biobank.com

populations of wild material, much in the way this selection was imposed during domestication, continues to be a major strategy for identifying improved plant cultivars for specific environments and uses. Plant breeding by cross pollination of plants became a widespread approach to developing superior crop plants over the last 100 years. Mutation breeding has also been important in many species as a mechanism for the introduction of a new genetic variation for traits of commercial importance. The discovery of the chemical basis of heredity with the determination of the structure of DNA more than 50 years ago has led to development of techniques for selecting plants for desirable traits by direct analysis of the DNA. The recent advances of DNA analysis technology has resulted in the complete genetic code of crop species being determined. The science of genomics (the analysis of all the genes in an organism) has begun to impact on agriculture and food production (Henry, 2009). Genomics reverses the previous paradigm in which the analysis traditionally targeted the discovery of genes to match a single specific trait. Now we can discover all the genes first and then ask what they do without necessarily having a specific objective or target. This has dramatically accelerated the rate of growth in genetic knowledge, providing a basis for more rapid and radical genetic improvement in crop species. Very recently recombinant DNA technology has become an option for introduction of novel traits. However, the major impact of DNA technology has been the more precise conventional selection of superior plants that DNA analysis allows. DNA sequencing techniques (the methods used to determine the genetic code of a plant or other organism) have been dramatically improved in the last few years. This technology is continuing to be improved in efficiency and effectiveness, and the full impact of its widespread application to plants is only likely to become apparent over the next couple of decades. Hybrid plants offer the potential for further improvements in plant productivity, and the rate of development and quality of hybrids may be greatly enhanced using the tools of genomics. Hybrid plants with higher yields due to the 'hybrid vigour' that comes from crossing two individuals that are genetically diverse may be

Box 2.6 *The Green Revolution*

The Green Revolution produced new cultivars of the major crops, increasing world food production to keep pace with the growth in world population in the period since the 1950s. Critics of the Green Revolution have charged that the new cultivars made farmers more dependent on high-cost inputs of fertilizer. While the full advantages of the new cultivars did require more nutrient input to deliver the full benefit, the new cultivars performed much better than the cultivars they replaced even in the absence of greater inputs. The simple laws of mass balance require that the removal of much greater amounts of crop as grain requires greater inputs to satisfy the need for replacement of soil nutrients. Norman Borlaug won the Nobel Peace Prize in 1970 for his contribution to saving the lives of millions with these advances in cereal breeding. A major achievement of the Green Revolution repeated in several major species has been an improvement in harvest index. Harvest index is the ratio of harvested crop to total plant biomass. Increased harvest index results in more grain harvest for the same total amount of plant growth. Borlaug's selection for a single gene, the semi-dwarfing gene in wheat, increased the harvest index of wheat, and is estimated to have provided the technology to allow the production of food for an additional 1 billion humans from the same area of land. Similar approaches have advanced the production of rice. Modern cultivars have a harvest index that allows about half of the total mass of the plant to be recovered as grain.

The implications of the Green Revolution for biodiversity will be discussed later in this book (Chapter 9).

considered the opposite of the poor performing individuals that result from inbreeding between closely related individuals. This technology is still to be perfected in many important food crops. These scientific advances may also provide new tools to domesticate additional food crops, which will be discussed in Chapter 10.

The potential to domesticate new species to expand the diversity and improve the security of food supply is generally viewed as being very limited. Diamond (2005a) has argued strongly that humans have domesticated most species with potential. Three types of exception to this may be found in:

1 species that are found in regions that had very few species that were suitable for domestication. A critical mass of species may be necessary to justify the transition to an agricultural lifestyle in any region;
2 species with new non-traditional uses and for which there was previously no reason for domestication (energy crops for biofuels may be a good example of this category); and
3 species that have barriers to domestication that can only be overcome with new technology that was not available in the past. Examples might include toxic plants that can now be screened for lower levels of toxin with modern analytical chemistry approaches.

Food from plants in the future

Humans have been very successful in domesticating plants for food and other uses. Plants have been subjected to genetic selection and deliberate cross breeding over a period of more than 10,000 years, resulting in highly domesticated plants adapted to human use. The challenge of the future will be to ensure that the full diversity of both the domesticated and wild gene pools is retained (especially for the major food crop species) to support food security by coping with the impact of the climate change (explored in the next chapter), and expansion of agricultural production to keep pace with the growth in human demand. The Consultative Group on International Agricultural Research (CGIAR) relevant to plant genetic resources (Table 2.6) plays a key role, especially in ensuring as far as possible ongoing outcomes of plant breeding of the staple food species are delivered to the poor in many countries.

Improving the nutritional value of food plants

We need to produce more food to satisfy growing demand, but we also need to produce food with an improved nutritional value – foods that improve human health are required. Our knowledge of links between diet and health

Table 2.6 *Consultative Group on International Agricultural Research (CGIAR) Centres working on plant genetic resources*

Centre	Main location
International Centre for Wheat and Maize Improvement (CIMMYT)	Mexico
International Rice Research Institute (IRRI)	Philippines
International Centre for Agricultural Research in the Dry Areas (ICARDA)	Syria
International Centre for Agricultural Research in the Semi-Arid Tropics (ICRISAT)	India
International Potato Centre (CIP)	Peru
International Food Policy Research Institute (IFPRI)	US
International Centre for Tropical Agriculture (CIAT)	Colombia
International Institute of Tropical Agriculture (IITA)	Nigeria
International Livestock Research Institute (ILRA)	Kenya
Africa Rice Centre (WARDA)	Benin
Centre for International Forestry Research (CIFOR)	Indonesia
World Agroforestry Centre (ICRAF)	Kenya
Bioversity International	Italy

Source: (CGIAR) Centres working with plant genetic resources: www.cgiar.org/centers/index.html

continue to improve and a major challenge is how to develop new cultivars of food crops to deliver health benefits. We can attempt to breed health into attractive foods or we can attempt to make healthy food more attractive to humans. The relative effectiveness of these approaches will be case-specific. The food industry tends to add ingredients that add health or at least allow food labelling that suggests a health advantage. However, plant breeders may also have dramatic and effective success in making highly nutritious foods more acceptable or attractive to human tastes or preferences.

In recent years there have been periods when the prices of foods have risen strongly due to reduced stocks and apparent inability of growth in supply to keep up with growth in demand. Ongoing underlying factors contributing to lower production and upward pressure on food prices include:

- growth in human population;
- growth in per capita consumption, especially associated with changes in diet (e.g. more meat) resulting from global economic growth;

Box 2.7 *Improving the folate content of cereals*

Humans require a range of essential nutrients in their diet. These include substances that are essential for biochemical processes in the body but cannot be produced by human metabolism. Folate is an example of this type of substance. Different forms of folate (tetrahydrofolate) and glutaminated derivatives collectively provide the folate in human diets. These are water soluble vitamins in the B group (folate is sometimes defined as vitamin B_9). Plants produce these molecules especially in green tissues. Vegetables (especially green leafy vegetables) and fruits are good sources of folate. Cereals and tubers are relatively poor sources of folate, but because of their large contribution to the diets of many humans these provide a significant part of the folate in many human diets.

Deficiency in folate is common in human populations both in the developed and developing world. A lack of folate has many serious consequences for humans – neural development and function are damaged by a lack of folate. This can result in congenital deformities such as spina bifida and anencephaly, and may contribute to dementia and Alzheimer's disease.

Folate biosynthesis in plants is a complex process involving metabolism in several different sub-cellular compartments, the cytoplasm, chloroplast and mitochondria. This complexity partially explains the lack of progress in developing plants with increased folate content.

Recently the genes in the folate biosynthesis pathway in wheat were characterized (McIntosh and Henry, 2008). This research indicated that folate was produced at all stages of the life cycle of the wheat plant, confirming the essential role of folate in the cells of all higher organisms. Even the dormant seed was able to produce folate. This knowledge has provided new tools for use in selecting cereals with elevated folate content.

This approach offers the potential to target the development of foods produced from plants with optimal nutrient content. Research has already defined many new options for plant breeders to improve human nutrition. We can expect science to deliver substantial further optimization of the nutritional value of major food plants. This is largely required because of likely continued failure of human populations to access a balanced diet by consuming an appropriate range of foods. The reason for this lack of balance in human diets varies in different populations and communities. Food preferences may be economic or cultural. However, improving the aesthetic value and the taste of nutritious foods remains very important. We need high folate content in foods humans prefer to eat.

- continuing removal of agricultural land by conversion to non-agricultural use (e.g. roads or urbanization);
- loss of land to desertification and increased salinity;
- increasing difficulty in continuing to find more water as production expands;
- competition for land and water with growing areas of energy crops, especially in the future as energy costs grow;
- loss of productivity in some regions due to climate change, especially in the future as impacts increase.

Despite obesity being a major world problem, food prices were historically high (probably for several complex reasons) and almost 1 billion people were suffering severe food deficiencies in late-2008 (FAO, 2008). The strong imperative to deal with these issues and ensure global food security adds to the risks that an expanding agricultural production footprint may have adverse impacts on biodiversity. There is an urgent need for a greater focus on research to enable more sustainable agriculture, and the development and implementation of policies to support sustainable agriculture and biodiversity conservation. These issues are made more urgent by the threat of climate change. The potential impact of climate change on food production will be discussed in the next chapter.

3

Impact of Climate Change on Food and Fibre Production

What's more, climate change is a breaking story. Just over thirty years ago experts were at loggerheads about whether Earth was warming or cooling – unable to decide whether an ice-house or a greenhouse future was on the way. By 1975, however, the first sophisticated computer models were suggesting that a doubling of carbon dioxide (CO_2) in the atmosphere would lead to an increase in global temperature of around 3°C. Still, concern among both scientists and the community was not significant. There was a brief period of optimism when some researchers believed that the extra CO_2 in the atmosphere would fertilise the world's croplands and produce a bonanza for farmers.

Tim Flannery, *The Weather Makers*

Global warming is now a widespread concern in the community. Climate change threatens crop and food production by changing the environmental conditions in traditional crop production areas. With supply and demand finely balanced, even small changes in climate over a few years become critical, especially as the scale of crop production required to support human populations increases. Climate change has been linked to increased levels of greenhouse gases such as carbon dioxide (CO_2) in the atmosphere. Recent research is firming up the causative link between greenhouse gas increases, warming and biological impacts (Rosenzweig et al, 2008). Climate change or even short-term variations in climate may have serious consequences for global food production when global food supply and demand are as evenly balanced as has been the case in recent years. Regardless of cause, climate variation is now a major risk to food security.

Carbon dioxide concentrations in the atmosphere

Historical analysis of atmospheric CO_2 concentrations has been followed back 650,000 years by examining bubbles trapped in ice in Antarctica, confirming a strong correlation between CO_2 and temperature (Luthi et al, 2008). CO_2 levels have been monitored for some time and show a steady increase from

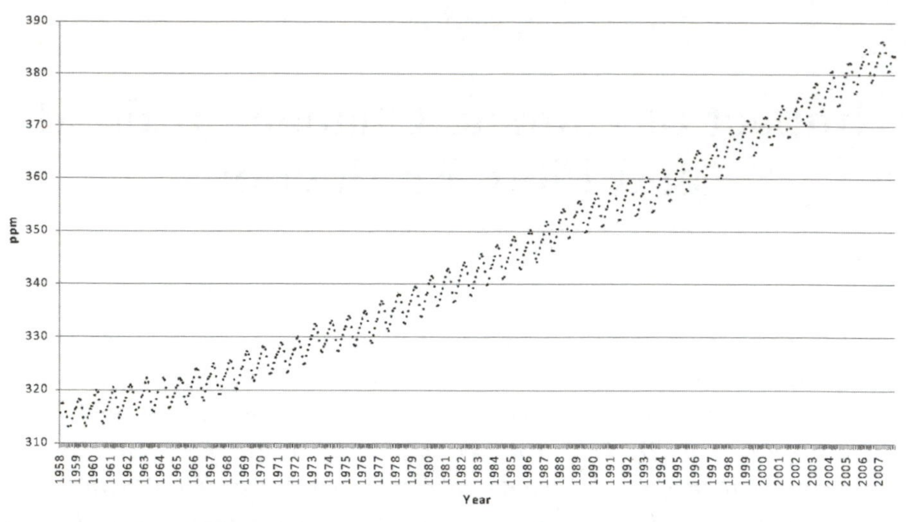

Keeling and Whorf, 2005

Figure 3.1 *Carbon dioxide concentrations in the atmosphere*

320ppm in 1960 to almost 380ppm in 2004 (Figure 3.1). The concentration of CO_2 in the atmosphere is currently rising at around 2ppm per year.

The concentration decreases each year in the northern hemisphere spring as trees grow new leaves, and increases again in the autumn as they drop their leaves and return carbon dioxide to the atmosphere. The dominance of the northern hemisphere forests is illustrated by this distinct pattern that can be detected worldwide. The underlying concentration increases each year despite this annual cycle. The fixation of CO_2 by plants also varies dramatically throughout the day as the plants respond to the daily cycle of light and dark, using the light energy to capture CO_2 to produce carbohydrates (Darbyshire et al, 1979).

Other greenhouse gases

Nitrous oxide (N_2O) is a potent greenhouse gas estimated to be 270 times as effective as CO_2, while methane is 21 times as effective. While these gases are far less abundant (much lower concentrations in the atmosphere) they can account for a significant part of the global warming impact because of their much greater influence. Some of the more unpredictable aspects of climate change may be associated with greenhouse gases other than CO_2. For example, recent research has identified an unexpected burst of methane from the tundra of Greenland at the point of freezing at the beginning of the

Box 3.1 *Scientific and popular views of climate change*

The impact of human activities on climate and the association between greenhouse gas levels and climate has been a major scientific and public controversy of the last couple of decades. The link between the increase in CO_2 and global temperatures has not been so easy for some scientists and the general community to accept. The acceptance of human-induced climate change has dramatic consequences. Scientists, by their nature and training, are sceptical of everything. The weather varies greatly and in the short term provides no real evidence of a permanent change in climate. The steady and continuing increase in greenhouse gases, especially CO_2, has been more convincing and is now widely accepted. CO_2 can be measured easily and objectively and the upward trend in concentration is indisputable. However, convincing data showing historical evidence for a very close association between CO_2 and temperature suggest one of only two possibilities: either CO_2 increases cause global warming or global warming causes CO_2 rises. The later option leaves the cause of global warming unanswered. Disturbingly the CO_2 concentration in the atmosphere have recently been rising more rapidly than that assumed in the models, providing some of the more pessimistic predictions of global impact.

winter (Mastepanov et al, 2008). Many aspects of the cycling of these gases remain to be discovered and explained.

Global temperature

Recent increases in global temperature are depicted in Figure 3.2. These values are relative to the mean temperatures in the period 1951–80 (Goddard Institute for Space Studies, NASA, 2008). Projected global temperatures in the future vary widely with the uncertainty of the likely changes in the atmosphere due to human activities and the difficulty of predicting the impact of this on global climate. Complex interactions may produce a wide range of different outcomes in specific locations. A more variable climate with a greater frequency of extreme weather events is widely predicted.

Climate Change reported in 2007 that the increase in global temperature was expected to have a number of impacts:

- altered weather patterns with more extreme events;
- rising sea levels;

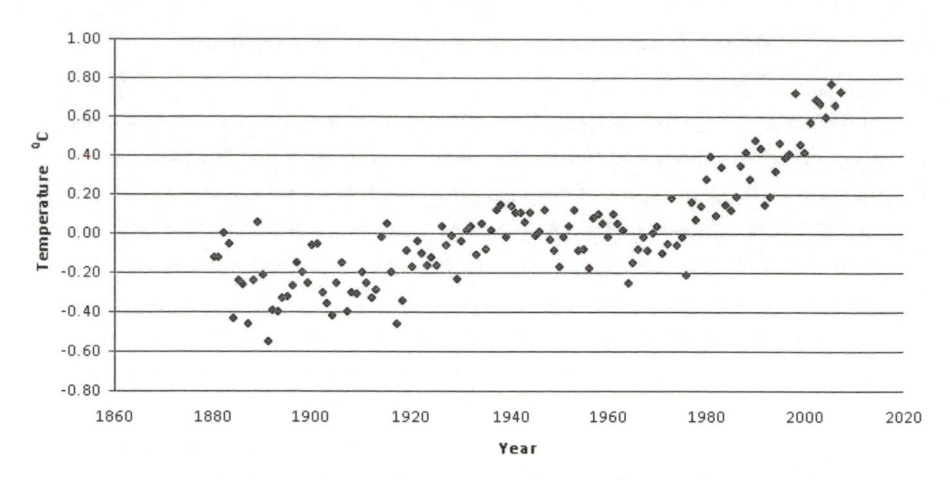

Figure 3.2 *Global temperature*

- changes in the biotic (living) environment (populations and distribution of micro-organisms and insects). The changed environment will impact on plant and animal disease incidence;
- changes in the availability of fresh water;
- loss of social and urban infrastructure (e.g. buildings, sewage and public service infrastructure).

Global temperature trends may also be impacted by other factors acting on different timescales. Global temperature will be the net result of all factors: those that are associated with greenhouse gas levels and those that are not. A recent analysis has modelled the temperature to the end of the century and predicted increases in temperature that would directly reduce crop yields and global food production (Battisti and Naylor, 2009).

Coping with climate change in agricultural production

The causes of climate change or more specifically global warming do not alter the risks to agriculture that is posed, but may influence the options available to deal with the problem. Higher temperatures and less rainfall (predicted impacts of climate trends in many areas) will reduce crop production potential. Relocating agricultural production to more favourable environments will offer at best a very limited option to cope with this. Greater environmental variation has also been predicted for many areas. This will increase the difficulty of managing agricultural production inputs to maximize productivity. Plants are now growing in an atmosphere with a CO_2 concentration much higher than the one to which they are adapted. The concentration of CO_2 in the atmosphere in 2007 averaged 383ppm (World Meteorological Organization, 2008). The photosynthetic apparatus of plants evolved in an atmosphere with a concentration of CO_2 in the 200–300ppm range. It may eventually be possible or necessary to re-engineer plants to perform better at higher concentrations of CO_2. The targets for these efforts will be defined by our ability to stabilize CO_2 concentrations at some reasonable level.

Plants have a diversity of photosynthetic pathways (Box 3.2). CO_2 is captured in most plants in a reaction that produces a three-carbon compound as the first product of photosynthesis. Some plants have evolved an additional mechanism for concentrating CO_2 inside the plant in high-light tropical environments that produce an initial four-carbon product. These C4 plants are able to perform better in these high-light environments fixing carbon more efficiently in hotter and drier environments. Rising concentrations of CO_2 in the atmosphere were originally thought to be likely to lead to a reduction in the advantage of C4 plants. However, while the direct impact of higher CO_2 concentrations is to reduce the advantage of C4 plants, the associated increases in temperature and reductions in water supply in the environment should ensure that C4 plants continue to perform better. Selection of C4 plants may become an important option in the adaptation of crops to climate change.

Box 3.2 *Pathways of photosynthesis*

All plants have the biochemistry to fix carbon by combining CO_2 with a five-carbon compound (ribulose 1,5-bisphosphate) to form two three-carbon molecules (3-phosphoglycerate). This is the essential feature of C3 photosynthesis.

Plants from warmer climates have developed an additional ability to first fix CO_2 by combining it with a three-carbon compound (phosphoenolpyruvate) to form a four-carbon compound (oxaloacetate). This is C4 photosynthesis. The passage of CO_2 into the leaf is associated with loss of water since CO_2 and water are molecules of similar size. Specialized anatomy in C4 plants allows these two processes to be separated. The four-carbon intermediate is transported to specialized cells, where it is de-carboxylated to produce bicarbonate that is then fixed in the normal C3 pathway. This effectively allows CO_2 to be concentrated in the leaf, allowing more efficient photosynthesis at high temperatures and with less water loss.

There are several variations on the C4 pathway involving different intermediates (e.g. malate and aspartate) (Vermerris, 2008a).

Adaptation of food and energy production to climate change may require the selection or breeding of more C4 plants.

Impact of climate change on wild crop relatives

The selection or breeding of crops better adapted to altered climates will be difficult at the speed that may be necessary to respond to this challenge. Climate change also threatens the survival of wild crop relatives, key long-term resources for crop adaptation and food security.

Recent research has examined the influence of climate on genetic diversity in wild populations of the first domesticated plant, barley (Cronin et al, 2007). Genetic diversity was found to be greatest at the driest sites and lowest in the wet or more favourable environments. The more extreme environments are probably the most variable and this may explain the greater genetic diversity that has evolved in the populations from the extreme environments to allow the plants to adapt to environmental stresses. Disease-resistant genes in these populations were found to be most diverse, suggesting that adaptation to dry environments may require adaptation to a different spectrum of pests and diseases in these environments. This is likely to be a key factor in adapting plants to climate change. The plant not only needs to cope with hotter or drier conditions, but also a new and possibly wider range of diseases. More details on this research on gene diversity in relation to the climate are given in Box 3.3.

Box 3.3 *Gene diversity in relation to climate*

An example of research on variation in a gene in wild barley is described here in relation to adapting food crops to climate change and in Box 8.1 in relation to biodiversity conservation. The variation in the sequence of the *Isa* gene and other genes in wild barley has been explored in relation to climatic variables. The research initially analysed the variation in the DNA sequence of a gene that had been characterized for some time. The original interest in this gene was because it encodes a protein (BASI) in the seed that inhibits the amylase enzymes that breakdown starch in the seed during germination. This was of great interest to wheat breeders attempting the breeding of wheat with resistance to pre-harvest sprouting (germination). The level of this protein had the potential to moderate the impact of rain damage on wheat bread-making quality. Rain causes premature germination of wheat in the field before harvest, resulting in the production of amylase that degrades starch during dough mixing. The sequence of the gene was analysed in the laboratory of Kihoharu Oono in the National Institute of Agrobiological Resources in Japan in 1990 (Henry and Oono, 1991). At this time the protein was thought to be found in the endosperm (the main starchy part of the seed). Much later, the work of a

PhD student, Agnelo Furtado, in collaboration with Ken Scott (University of Queensland) (Furtado et al, 2003), revealed that the protein was unexpectedly in the outer parts of the seed (pericarp). This discovery suggested that the protein had a role in defence of the seed against pests or diseases. Collaboration with Eviatar Nevo, Institute of Evolution, Israel, allowed an examination of the diversity of this gene in wild barley populations in relation to environmental variation (Cronin et al, 2007). More recently this work has been extended to the study of other genes in these same populations. These have included abiotic stress-related genes such as betaine aldehyde dehydrogenase and alcohol dehydrogenase, and biotic stress resistance genes that have been associated with resistance to fungal diseases. Experiments to test the variation of these genes in wild populations of other species followed. This research showed great variation in gene diversity between populations that may be associated with adaptation to climate in wild plant populations. Understanding these processes will be important if we are to rapidly adapt crop plants to climate change.

Impact of climate change on food production

Climate change is likely to have a major impact on the production of crops. The level and type of impact is expected to vary greatly in different locations. A regional or even local analysis is required to fully understand the implications of climate change. Cline (2007) has estimated the impact of global warming on a country and regional basis.

Africa

Much of Africa is already marginal for agriculture with high temperatures and limited water supply. Drier conditions over much of Africa are likely to result in more frequent crop failures. Agricultural output has been estimated to decline by 17–28 per cent by the 2080s. The availability of water for irrigation will be critical. The supply of water from the Nile will be essential to allow continued production in Egypt. Dryland agriculture in Africa is expected to suffer substantial adverse impacts from climate change. Climate change in Africa is likely to have overall strong negative impacts on people that are probably among those less able to cope with reductions in income or food supply. The likely impact and the need for efforts to adapt to climate change in Africa have been analysed by Dinar et al (2008).

Asia

The rapid industrial and economic development of Asia makes this region a major and growing contributor to greenhouse gas emission. Asia accounts for more than 60 per cent of the world population and food production is likely to be impacted by reduced availability of water in many areas. Output has been estimated to decline by 7–19 per cent by the 2080s. India is projected to suffer significant losses of production ranging from 30–35 per cent in the south to 60 per cent in the north.

Europe

Increased temperatures may improve crop yields in northern Europe, but this is likely to be offset by negative impacts in southern Europe. Grain production could be enhanced in areas (e.g. Norway and Finland) that are currently limited by low temperatures, but warmer parts of Europe, such as Spain and Italy, are likely to suffer reduced capacity.

North America

Production in Canada may expand north but reductions in the midwest of the US are likely due to reductions in rainfall. The overall situation in the US is relatively balanced to slightly negative. Southern areas will face reduced productivity because of excessive temperatures: Mexico is expected to suffer the greatest losses in production.

South America

The impact of climate change in South America is strongly negative, with greater losses in Brazil than in Argentina. Brazil is currently a major food producer with potential for significant growth in agriculture if climate does not deteriorate. Again, areas such as Brazil, that are already warmer, have the most to lose as climates become too hot for agricultural production.

Australia

Australia is one of the developed countries with a significant potential for a loss of around 16 per cent in agricultural productivity by the 2080s (Cline, 2007). Temperatures are already above the optimal level in many areas. Water is in limited supply, preventing irrigation to cope with the increased water needs of higher temperatures. Much of Australia has low and variable rainfall. Southern Australia has a Mediterranean climate with rain mainly in the winter and a hot dry summer. Northern Australia has more monsoonal

Box 3.4 *Case study – cereals (rice, wheat, barley and sorghum) in Australia*

Australia is a country with a relatively dry climate under threat of further reductions in rainfall associated with global warming. Rice has for many years been grown in low rainfall inland areas of southern Australia under irrigation. Rice production has been reduced dramatically in the last few years as irrigation water has not been available due to prolonged drought. Perennial crops such as grapes and tree crops have expanded and gain priority in access to irrigation water relative to rice where the decision to plant can always be delayed another year. A return to the peak levels of production in the Australian industry may not now be possible because of competition from more permanent crops, and a probable loss of infrastructure and capability if low production persists. Rice production in northern Australia under rain feed conditions might be an option for the future, especially as a response to global warming. Rice crops were grown on the east coast of Australia in 2008. A crop in northern New South Wales was the first commercial scale crop of rice in this region.

Wheat and barley are grown as winter crops in Australia, predominantly across southern Australia in Mediterranean climates. Australia produces only a modest amount of wheat, but is a major exporter because of low population and resulting low domestic demand. Australia is a major exporter of barley and with Canada accounts for most of the traded barley in the world. Exports of both wheat and barley were greatly reduced in 2007/08 because of low production due to dry conditions across southern Australia. Barley is more tolerant of drought stress than wheat and frequent dry seasons could increase the proportion of barley planted. The current reductions in production are expected to be short term, reflecting seasonal conditions, but indicate the likely impact long term of climate change. The low production of the past year may become more normal. Sorghum production has increased greatly in recent years in the northern Australian grain regions on the east coast. Sorghum is grown opportunistically on summer rainfall and as a result climate change may impact very differently on this crop.

Overall cereal production in Australia may change in response to global warming, with a reduction in total production but with the possibility of a move of rice production to the north, an expansion of sorghum production and a reduction in barley, but probably more especially wheat production in the main production areas in the south.

rainfall mainly in the summer with warm dry winters. The variation in rainfall between seasons has been associated with climatic cycles over the Pacific Ocean. The El Nino pattern associated with dry conditions in Australia has become frequent recently. The main inland river system (Murray/Darling) has had record low inflows in recent years.

Genetic strategies for adapting food production to climate change

The above analysis of possible impacts on cereal production in Australia assumes no change in the ability of these species to perform in more extreme environments. As climate change impacts, great efforts will be made (and are already being made) to adapt these key species to cope with their new environments. However, the potential to breed better adapted cultivars quickly enough to cope with the pace of climate change is unlikely. Future needs include many options that are likely to be longer-term outcomes of current research activities. These include novel strategies for disease and drought tolerance, nitrogen use efficiency and more efficient photosynthesis. Moving production of each species to more suitable environments will be a key strategy in the shorter term. Changing species will be another main response. The more stress-tolerant species such as barley and sorghum production may increase at the expense of wheat and rice. Unfortunately wheat and rice are more attractive human foods. Developing food technology and genetic selection of better food cultivars for these more adaptable species should be a high priority.

The magnitude of climate change that is now expected will be a major challenge for plant breeders. Adaption of plants rapidly enough to maintain current production levels of major food crops long term will probably require significant research investment and probably represents an ambitious target. The predicted growth in demand will not necessarily be met in this scenario, but options to limit growth in demand are not easy to identify.

Agriculture needs to become part of the solution to climate change, not just a victim. Research on how food production can contribute to reductions in greenhouse gas production is an important area for investment (IFPRI, 2009). Innovations in agriculture to support mitigation of climate change need support.

We will return to a discussion of the potential for future plant production to satisfy both food and energy demands in Chapter 6, but first we will consider human energy needs (Chapter 4) and the potential contribution of plants (Chapter 5).

4

Energy Resources

Petroleum-based fuels and related materials are central to the economies of developed and developing countries around the world. However, these resources are finite and expected to enter a period of diminishing availability within the next several decades.

Ahmann and Dorgan (2007)

Human societies use energy in many different ways, in transportation, and in industrial and domestic applications. Energy can be sourced from solar, wind, wave, hydroelectric, geothermal and nuclear sources. Traditionally humans have used plants as a direct source of biomass for heating and cooking. However, more recently, fossil fuels (coal, gas and oil) have been used as energy sources. These reservoirs of carbon compounds produced from CO_2 by plants growing over long periods of time have been used extensively to produce electricity and liquid fuels for transportation. These resources are also used as chemical feedstocks for the manufacture of a wide range of carbon compounds from plastics to pharmaceuticals. The burning of fossil fuels results in an increase in greenhouse gases, especially CO_2 and the associated threat of global warming. Human societies have a great dependence on oil and for transportation. Oil stocks are declining and prices of products derived from oil have been increasing sharply in recent years. It has been estimated that humans have consumed around 875 billion barrels of oil since we started using oil, that about 1.7 trillion barrels remain in established reserves and that we might eventually find another 900 billion barrels (Ahmann and Dorgan, 2007). Two main factors encourage the search for alternative sources of energy. Firstly, the growing cost of oil and in the long term the ultimately limited nature of the resource makes alternatives attractive. Secondly, the risks of global warming associated with the consumption of fossil fuels provides further incentive to develop new technologies. Options that replace oil for high volume uses such as transportation may also help conserve oil stocks for more critical or less substitutable uses such as chemical feedstocks. Photosynthesis by plants captures light energy by using it to combine CO_2 to form carbon-containing compounds mainly in the form of carbohydrates. Simple calculations indicate that the amount of energy captured by photosynthesis each day is far greater than the amount of energy used by human societies. In 2008, for the first time, the International Energy Agency (IEA, World Energy Outlook, 2008) called for an urgent effort to move away from oil to a more sustainable energy supply system to avoid catastrophic climate change.

Growth in human energy consumption

Human consumption of energy is growing at about 2 per cent per annum, with almost twice this growth rate in Asia. Consumption grew by an estimated 2.4 per cent in 2007, with China accounting for half of this global growth (BP Statistical Review, 2008). Natural gas consumption continues to grow more rapidly than oil consumption. Coal consumption showed even stronger growth at 4.5 per cent. Wind and solar power growth is rapid at 28.5 per cent and 37 per cent respectively for 2007. Nuclear power generation actually fell (due to an earthquake) and hydroelectric power generation grew by 1.7 per cent in 2007. Around 80 per cent of the energy consumed is currently from fossil fuels. Total energy traded in the world in 2007 was around 11,000 million tonnes (oil equivalent). Growth in oil consumption is depicted in Figures 4.1 and 4.2.

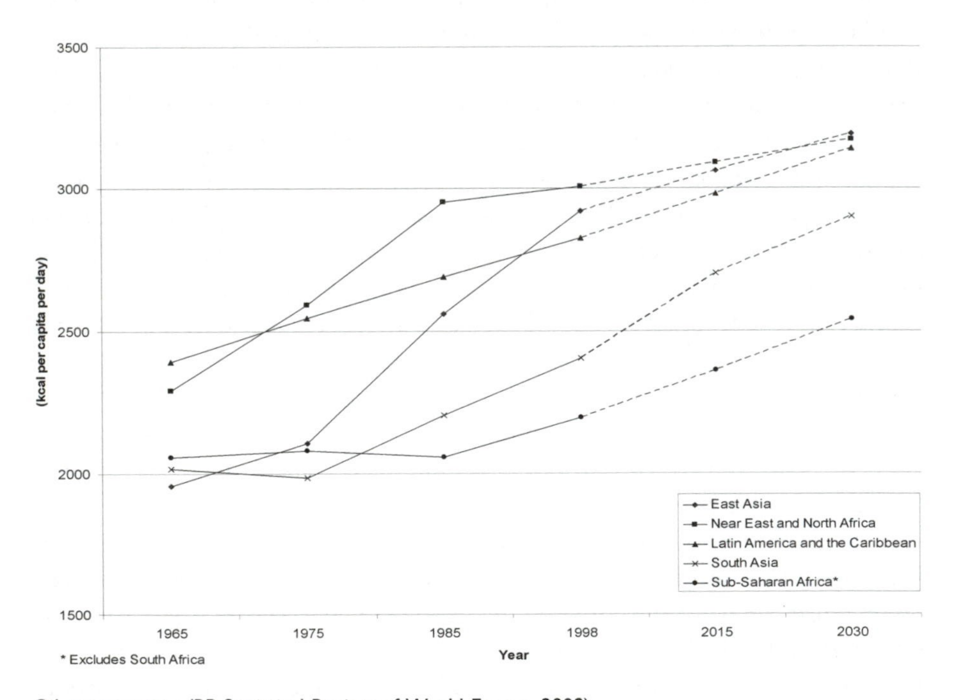

Oil consumption (BP Statistical Review of World Energy, 2008)

Figure 4.1 *Oil consumption – regions*

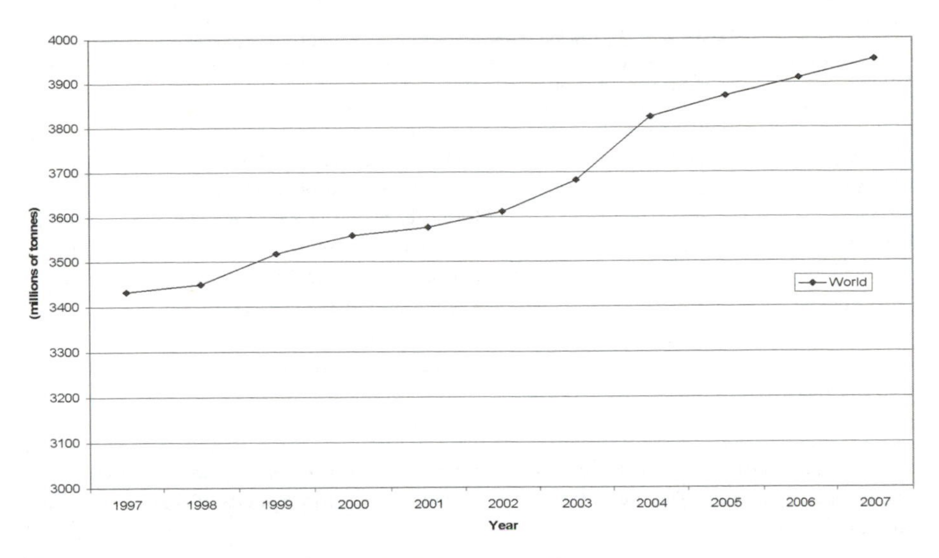

Oil consumption (BP Statistical Review of World Energy, 2008)

Figure 4.2 *Oil consumption – world*

Non-transport uses

Electricity generation is widely based upon the burning of fossil fuels such as coal. Alternative options (Table 4.1) such as nuclear power are now being considered more seriously because of concerns about greenhouse gas emissions. Methods of capturing CO_2 in coal-powered power stations are being investigated. The success or otherwise of this technology may determine the future of electricity generation from fossil fuels.

Transport use

Common transport fuels include petroleum (gasoline or petrol and diesel) for use in cars and trucks and specialty fuels for aeroplanes. Unless non-

Box 4.1 *Approximate conversions for units of energy*

1 tonne of oil = 1.16 kilolitres of oil = 7.33 barrels of oil = 307 US gallons of oil

1 tonne of oil = 10 kilocalories = 42 gigajoules = 40 million BTU (British Thermal Units) = 1.5 tonnes of hard coal = 12 megawatt-hours of electricity

Source: BP Statistical Review of World Energy, 2008

Table 4.1 *Electricity generation from non-fossil fuel sources*

Source	Current capacity	
	(gigawatts)	
Hydro	800	45,000 large dams globally
Nuclear	370	439 operational nuclear power plants
Wind	94	Capacity growing 25% per year (last 5 years)
Biomass	40	Mostly used directly for cooking and heating. Specially suited to liquid (biofuel) production
Geothermal	10	High-grade resources are rare
Solar	9	Only during daylight. Limited by storage technology
Oceans (waves and tides)	<1	

Source: Schiermeier et al (2008). Electricity accounts for around 40 per cent of total energy consumption by humans. The use of fossil fuels to produce electricity releases 10 gigatonnes of CO_2 per year.

carbon-based fuels such as hydrogen are developed or electric cars are adopted, carbon-based biofuel production will remain an important option to replace fossil fuels for transport. A major advantage of carbon-based fuels is that they are very energy dense (currently 10–100 times more dense) compared to energy stored in the most efficient batteries. Energy density (the amount of energy per unit of weight) is a key characteristic for a fuel to be used to power a mobile vehicle.

'Renewable energy'

The use of renewable sources of energy is an attractive alternative to the consumption of fossil fuels. Solar and wind power are widely considered renewable energy sources. They do depend on energy from the Sun which is ultimately not renewable, but on the timescale of human lives these sources of energy are effectively available forever. Geothermal energy may be harvested by circulating water underground to be heated and returning the water and the energy captured to the surface. This energy is also not strictly renewable, but may have major advantages in avoiding the release of greenhouse gases associated with fossil fuel consumption.

Bioenergy

Plants capture energy directly from the Sun in photosynthesis. Fossil fuels are derived from ancient plant material that has been accumulated in deposits that can be mined. The problem with the use of fossil fuels is that we are rapidly returning to the atmosphere large quantities of CO_2 that have been stored underground over very large periods of time. Bioenergy can be produced by growing plants and using them directly in the generation of electricity, production of fuels or chemical feedstocks. Biofuels is the term now used most specifically for transport fuels (currently usually ethanol but also biodiesel) produced from plants. This approach may reduce the greenhouse gas emission associated with the use of transport fuels. Carbon dioxide captured by the plant as it is grown is released back into the atmosphere when the fuel is used. This recycling of carbon dioxide avoids a net increase in CO_2 due to the release of carbon trapped in fossil fuels into the atmosphere. The extent to which this benefit is achieved depends upon the efficiency with which the plant can be grown and converted to a biofuel. Energy consumption and associated CO_2 emissions during the whole cycle of production need to be considered. For example, the energy used in preparing the field, planting the crop, applying fertilizer, making the fertilizer, harvesting the crop, transporting the crop to a processing plant, conversion to biofuel and transport of the biofuel to a retail outlet all need to be determined to establish the relative energy and greenhouse gas impact of this technology. Biofuels that generate more energy consumption than they produce are possible and probably the easiest type of biofuels to achieve in a technical sense. Biofuels with a more desirable environmental impact are needed and much current research and development effort is being devoted to making biofuels more energy efficient and environmentally desirable. We are already seeing dramatic gains in efficiencies of commercial biofuel facilities.

Research needs for biofuel production from plants

The development of technology for the economic and environmentally friendly biofuel production from plants requires significant research and development. The risk for those making this substantial investment is that another technology, for example, one that avoids the use of carbon fuels altogether (e.g. a very efficient technology producing hydrogen as a transport fuel directly from the Sun), could be developed and make this technology redundant. This is probably a scenario that is highly desirable from an environmental perspective; however, we cannot predict how long it might take and so we are faced with an urgent need to improve current biofuel technologies at least for the short to medium term. The production of biofuels from plants on a large scale has other risks associated with demanding more of our agricultural production systems. Competition with food crop production may reduce food production and increase food prices. Expanded agricultural production may demand more land is cultivated, threatening biodiversity and nature conservation. This issue will be addressed later in this book, but before that we need to examine the use of plants for current biofuel technologies and their limitations and promises of efficient second and later generation biofuels that research and development might deliver. This will be the topic of the next chapter.

5

Plant Resources for Bio-energy and Chemical Feedstock Uses

The cell walls of vascular plants account for much of the carbon fixed during photosynthesis and make up much of their biomass.

Philip Harris (2005)

Biofuels from plants

Growing fresh plant biomass represents a source of carbon for energy and feedstock production that is an alternative to the use of deposits of ancient plant biomass (fossil fuels). Rather than return CO_2 to the atmosphere as we do when we burn fossil fuels we have the potential to recycle carbon, fixing it in growing the plant and returning it to the atmosphere when the biomass is consumed. Biofuels may be produced in a more carbon neutral process, avoiding the addition of greenhouse gases to the atmosphere and the associated risk of global warming. For example, life-cycle assessment of ethanol production from switchgrass has produced an estimate of a 94 per cent reduction in greenhouse gas emissions compared with conventional fuel from oil (Schmer et al, 2008). Despite this potential the current first-generation technologies for biofuel production (based upon conversion of non-structural carbohydrates (sugars and starch) to ethanol or plant oils to biodiesel) have been assessed as often having minimal advantages, and may in fact have a net negative impact when all social, environmental and economic factors are considered (Charles et al, 2007). The efficiency of these first-generation processes is being rapidly improved. However, current substantial investments in improvement of technologies for biofuel production (US Department of Energy, 2008) are essential to achieving the promised potential of plants to deliver significant reductions in greenhouse gas emissions.

Electricity from plants

Plants may be combusted to generate electricity directly. This approach has been adopted by the sugarcane industry with the widespread use of the fibre residue for electricity generation after extraction of the sugar. Initially the objective of these processes was to generate enough electricity to power the sugar mills, but the technology has been developed to allow electricity to be generated significantly in excess of these requirements and to be used for domestic and industrial applications. This may be the best option for use of some types of biomass until more effective technologies for biofuel production are developed. However, biofuels could become a preferred use if efficient technologies can be perfected. The first generation of biofuels has involved the production of ethanol from corn or sugarcane and biodiesel from oilseeds. Current technology developments aim to greatly increase the fuel value per hectare of crop produced relative to these first generation technologies by developing the technology needed to utilize the structural carbohydrates (cell walls) of plants. The composition of the biomass and the available technologies determines the suitability of biomass resources for these competing applications.

Composition of plants for energy production

Plants store carbon mainly in carbohydrates. The presence of structural carbohydrates such as cellulose in the cell walls of plants, and non-structural carbohydrates such as sugars and starch within the cell, was introduced in Chapter 2. The potential of plants to replace oil in fuel and chemical feedstocks relates directly to their carbohydrate content and our ability to convert these carbohydrates efficiently into fuels and chemical feedstocks.

Non-structural carbohydrates in plants

Plant cells store carbohydrates as sugars. In green plants, the disaccharide – sucrose – predominates, but the two monosaccharides that make up sucrose – glucose and fructose – are usually also present. Fungi tend to accumulate the disaccharide trehalose (two glucose units). All plants accumulate glucose, fructose and sucrose, but some plant groups also produce other sugars. Plants from the family Rosaceae (apples, apricots, cherries, peaches, pears and plums) accumulate a sugar alcohol – glucitol (the alcohol resulting from reduction of glucose) – often called sorbitol. Sorbitol is the main sugar in a mature apple. The amount of sugar that plant cells can accumulate is limited. The osmotic impact of very high sugar content limits accumulation. To allow accumulation of more carbohydrates in the cell plants polymerize simple sugars – polymers of glucose (starch) and fructose (fructans) are most common. Almost all plants produce starch which accumulates as starch granules. Starch is an effective carbon sink in the cell, but carbon in these large structures may not be available for immediate metabolic use. Many plants from cool or dry environments have also evolved the ability to accumulate fructans (see Chapter 11 for details of families of plants containing fructans). These soluble fructose polymers can be readily polymerized and depolymerized in the plant to allow physiological adaptation to cold or drought stress. Some plants accumulate galactose polymers based upon sucrose (the raffinose series oligosaccharides); in the human diet these are commonly encountered in beans.

The non-structural carbohydrate content varies greatly in higher plants and has enormous potential for modification to better suit the needs of biofuel production. Increased levels of sucrose or starch are important targets for crops such as sugarcane and maize that are currently grown as sources of these carbohydrates for biofuel production (Smith, 2008). Alteration in the composition and properties of plant starch for a wide range of applications has been the subject of considerable commercial interest (Waters and Henry, 2007).

Non-structural carbohydrates as a source of biofuel

These carbohydrates are very readily available for use in fermentation. Biofuels generated by fermentation of plant non-structural carbohydrates have been termed first-generation biofuels to distinguish them from those produced from structural carbohydrates (second-generation biofuels). Fermentation of plant carbohydrates to produce ethanol has been a technology long utilized by humans to produce alcoholic beverages. Beer and wine production probably date back to the beginnings of plant domestication and agriculture. The question of whether beer or bread came first illustrates the point. A mixture of ground cereal (flour) and water can become bread or beer, and early agricultural communities probably produced both relying as they did on some of the same core technologies. Adaption of these long-established 'biotechnologies' for the production of fuel ethanol from the non-structural carbohydrates of plants has been a relatively simple process, building upon the long human history of brewing, distilling and wine making. More recently bacteria have been developed as an alternative to yeast in fermentation of sugars to produce ethanol. Plants expressing starch-degrading enzymes would assist the conversion of starch to fuel. Maize expressing a heat stable amylase has been developed to facilitate efficient conversion of the starch to glucose during processing.

Structural carbohydrates in plants

A defining feature of plants is the presence of a cell wall. The cell walls are the 'skeleton' of the plant providing structural support to the plant. The plant cell is much like an animal cell inside a rigid box (the cell wall). Plant cell walls may be thought of as being like reinforced concrete with fibres of cellulose (the metal rods) surrounded by non-cellulosic polysaccharides (the concrete). This is further strengthened by the addition of lignin and phenolic cross links between the polysaccharides. Ester-linked ferulic acid, *p*-coumaric acid and lignin provide cross linking. Plants contain a diversity of non-cellulosic polysaccharides.

The cell walls of grasses contain polymers of glucose that are related in structure to cellulose, but contain some glucose resides linked via carbon 3 rather than carbon 4 as in cellulose. This changes the properties of the polymer from a linear and rigid molecule likely to associate with others to form long micro-fibrals (as in cellulose) to an irregular molecule that is not likely to polymer in solution (the viscous beta-glucan solutions). Linear polymers of xylose with arabinose resides attached (arabinoxylans) are also abundant in these plant cell walls. The cell walls of the Commelinoid associate with others, but to exist as a long group of families that includes the grass family (Poaceae) form a distinct group within the monocotyledonous plant families (Henry and Harris, 1997) sharing common cell wall structures (polysaccharide composition and the presence of the ferulic acid cross links). These type II cell walls contain higher amounts of cellulose and very little pectin or protein. These features distinguish this group of plants from other higher plants (Figure 5.1). Other species of higher plants have type I cell walls containing xyloglucan as the main non-cellulosic polysaccharide, together with pectic polysaccharides. These polymers have often been called hemicelluloses (half cellulose), but use of this term can lead to confusion because it has been variously used to include all non-cellulosic polysaccharides and alternatively to include only neutral polysaccharides and exclude pectin (polysaccharides containing acidic residues such as galacturonic acid).

The diversity of cell wall structure and a growing understanding of cell wall biosynthesis suggest that significant cell wall modification to suit biofuel production may be possible (Pauly and Keegstra, 2008).

The composition of plant biomass can be used to estimate ethanol yields in conventional conversion technologies (Table 5.1).

Box 5.1 *Advances in technologies for the analysis of plant carbohydrates*

The rate of progress in understanding the composition of plants and in selecting appropriate cultivars better suited for specific food or other uses has relied on the techniques we have for the analysis of the carbohydrate composition of plants.

The approaches to these analyses were initially based upon the chemical approaches that had been used in the early determination of the structure of carbohydrates and their chemical preparation and synthesis.

Determination of the monosaccharide (simple sugar) composition of a polysaccharide (a polymer composed of many monosaccharide units (simple sugar residues) linked together with covalent chemical bonds) has been based upon the breakdown of the polysaccharide into the monosaccharide units and their analysis, usually following separation of the component sugars by some form of chromatography. A common approach has been to use acid to break down the polysaccharide into monosaccharides. The sugars in the resulting hydrolysate are complex to analyse because each sugar (e.g. glucose) can exist in a range of chemical confirmations and ring structures (e.g. alpha and beta). This complexity has usually been minimized by chemically reducing these sugars (aldehydes) to alditols (sugar alcohols). In this process all forms of glucose become a single open-chain molecule – glucitol. The resulting sugar alcohols can then be analysed. This has often been achieved by acetylating the hydroxyl groups on the sugars to form volatile alditol acetates to allow separation and analysis by gas chromatography. The protocols for these analyses prior to around 1980 were very time consuming and used approaches based upon preparatory organic chemistry.

The author was involved in a series of collaborations in the laboratory of the late Professor Bruce Stone in the early 1980s, which developed a series of methods to progress approaches that were based much more on the strategies of analytical chemistry or biochemistry. These methods were widely adopted because of the much larger numbers of samples that could be routinely and quantitatively analysed, allowing the variation in carbohydrate composition in biological samples to be more widely explored. The first paper in a series published on these methods (Blakeney et al, 1983) has been cited in more than 1000 publications. The same approach was used to refine the techniques for methylation analysis of polysaccharides (the method used to determine how (through which carbon atoms) the monosaccharide resides were linked together in the polymer) (Harris et al, 1984). These methods have also been cited widely in the scientific literature.

These techniques are still relevant today, but can now be complemented by the use of more advanced instrumental methods of analysis. Further developments and the application of new tools to analysis of the carbohydrate composition of plant materials, especially complex aspects of macromolecular structure, would facilitate the accelerated development of plants as improved biomass for biofuel production.

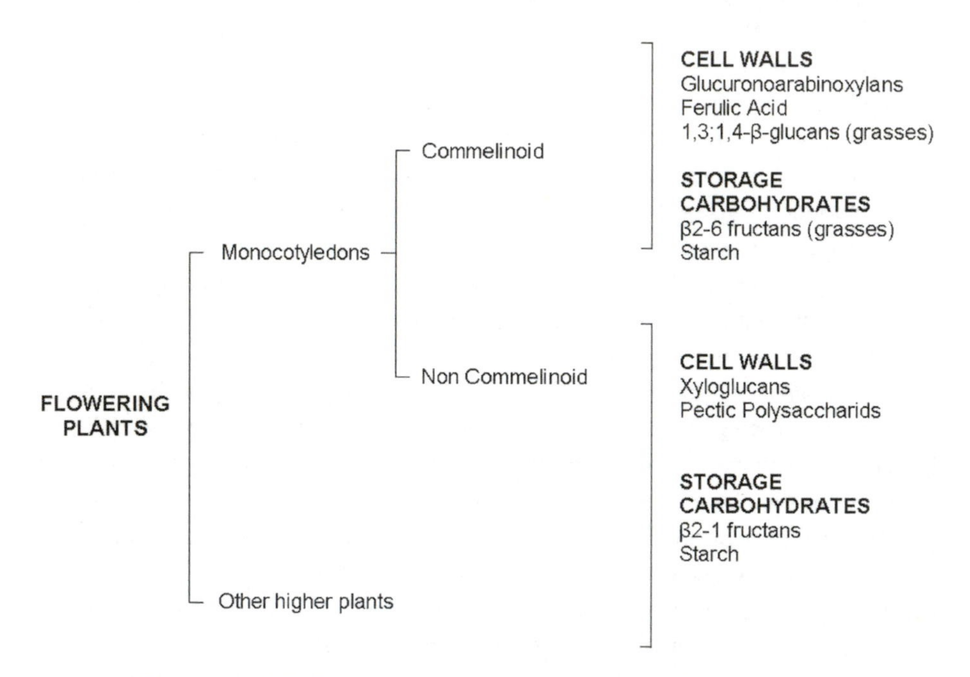

Figure 5.1 *Difference in biomass composition in flowering plants*

Structural carbohydrates as a source of biofuels

Structural polysaccharide is abundant in plant material and as such an attractive source of carbohydrate for conversion to fuels. This process is much more challenging technically than the conversion of sugars and starches. Cost-effective technologies for conversion of plant cell wall to biofuel are likely to be the key to the success of biofuels. Knowledge of the structure of the cell walls in specific groups of plants is important in developing technologies for efficient conversion of this material to biofuel. For example, the chemistry of the cell walls in grasses and related plants from the Commelinoid group (as defined above) may dictate different processing requirements compared to that required for other plant species.

Table 5.1 *Examples of the composition of plant biomass and predicted ethanol yields*

Biomass	Hexose content	Pentose content	Lignin content	Ethanol predicted yield (%)
Sugarcane (baggase, fibre residue after sugar extraction)	33	30	29	28
Wheat (straw)	30	24	18	24
Rice (hulls)	36	15	19	27
Pine	44	26	29	31
Eucalyptus (*E. saligna*)	45	12	25	25
Willow	37	23	21	27
Poplar	48	27	19	33
Newspaper	61	16	21	34

Source: Chandel et al, 2007

Biochemical conversion

Biochemical conversion involves hydrolysis of polysaccharides using acid or enzymes, followed by fermentation to produce fuel.

An important area of innovation in biochemical conversion technologies is in the pre-treatment of the biomass to improve the efficiency of the carbohydrate hydrolysis. The aim of pre-treatments is to change the biomass structure to make it more amenable to subsequent processing. Pre-treatments may include mechanical, thermal and chemical processes. Alkaline and treatments may also be included. The key objective of most research is to find a cost-effective pre-treatment method for the target biomass. Energy and greenhouse gas efficiency of pre-treatments are also important.

Acid hydrolysis may be a multi-step process with options for different acid concentrations and temperatures. Acid hydrolysis especially at high temperature can cause degradation of the monosaccharides to furans that may inhibit subsequent fermentation.

Enzymes that digest cell walls are known for microbial sources. Wood-rotting fungi have long been studied as a source of these enzymes. Micro-organisms from the gut of animals that derive energy from the cell walls of grasses or wood (e.g. ruminant animals such as cows and sheep, and insects such as termites) are another important source. These processes have long been possible, but achieving a high level of conversion at low cost, as is required for a commercial process, has proven more difficult. Combinations of acid and enzymes may be used to digest cellulosic biomass to produce sugars for fermentation to fuel molecules. Research aiming to produce these enzymes directly in the plant is in progress. This technology offers the potential to eliminate the cost of enzymes and to reduce the cost of mechanical and other processes required to allow access of the enzyme to carbohydrates within the plant material. Research focuses on producing more efficient enzymes at a lower cost and methods for pre-treatment of the plant material to allow the enzymes better access to the polysaccharides. Conventional micro-organisms used for fermentation cannot process both C6 and C5 sugars at the same time. The conversion of six carbon (C6) sugars such as glucose and five carbon (C5) sugars such as xylose is being tackled in many ways. The polysaccharide may be separated before hydrolysis to allow separate fermentation or organisms capable of simultaneous fermentation may be engineered. In other strategies, enzymes may be used to convert these sugars to forms better suited to simultaneous fermentation.

Thermochemical conversion

Thermochemical processes avoid some of the need to optimize the composition of biomass. The main disadvantage of these processes is that they generally require high-energy inputs. The production of biodiesel from cellulosic biomass by thermochemical methods is a process that was developed long ago, but is currently being re-examined and applied. Biomass can be heated in the absence of oxygen (fast pyrolysis) to directly produce an oil, or heated in the presence of a small amount of oxygen and steam (gasification) to produce a gas (syngas) that can be burnt to generate heat energy (e.g. for the generation of electricity) or subsequently converted to a liquid fuel (Fisher-Tropsch process). The use of biomass to generate syngas can result in high levels of ash which may damage turbines, indicating the need to develop sources of biomass with a low ash content. The development of improved catalysts is an active area of research and is likely to be a key factor in determining the success of thermochemical conversion. Pyrolysis oil may become 'biocrude oil', allowing the concentration of energy in biomass before transport to centralized refineries.

Chemical conversion

Chemical- rather than fermentation-based approaches may have advantages in being able to be readily scaled up. Recent novel strategies have been suggested for the production of high-value fuels from plants without fermentation or the use of micro-organisms. For example, 2,5-dimethylfuran (DMF), a desirable fuel molecule, could be produced by converting glucose to fructose with enzymes, acid-catalysed conversion of the fructose to 5-hydroxymethylfurfural (HMF) and the use of metal catalysts and hydrogen to produce the DMF (Schmidt and Dauenhauer, 2007).

Conversion technologies that combine one or more of these different technologies are also being developed. For example, biochemical methods may be used to convert the carbohydrates to sugars, while chemical methods may be applied to the lignin component. Separation into a low lignin fraction for pyrolysis and a high lignin component for gasification is another strategy.

Lignin

Lignin is a major component of plant biomass. Lignin may be viewed as the product of polymerization of three hydroxycinnamyl alcohol precursors, resulting in p-hydroxyphenyl, guaiacyl and syringyl units in the lignin. Biological degradation of lignin by enzymes and micro-organisms is difficult.

Plants with low lignin content are being selected for biofuel production to overcome the problems associated with lignin degradation. The brown midrib mutants of maize and sorghum have low lignin content and may allow improved biofuel yields from biomass of these species (Li et al, 2008). However, a high lignin is desirable in plant biomass that is combusted to produce heat energy or electricity. Recent reports also suggest that conversion of lignin into high-value alkanes (C8 and C9) may be possible at high yield using suitable catalysts. These developments might lead to an interest in developing high lignin biomass.

Oils

Plant oils are a more direct source of fuel molecules. Plants have been widely cultivated for their edible oils. Many of these can be used to produce fuel, but fuel production can extend to the use of species that have not been suitable for human food use because of the presence of toxic compounds in the oil. The yields of oil relative to biomass are generally very low, making the use of plant oils a poor option compared to biomass production when the required land, water and other resources are considered.

Types of biofuel that can be produced from plants

The type of biomass available influences the processes that are available to convert it to fuel and the types of fuel molecule that can be produced. Innovations in processing or conversion technology may allow more desirable fuels to be produced. This in turn may dictate different biomass specifications. This iterative interaction between developments in the target fuel, the conversion technology and the biomass is depicted in Figure 5.2 and Table 5.2. This has led to the use of terms such as first-, second- and later-generation biofuel to signify different combinations of biomass source, conversion technology and biofuel product.

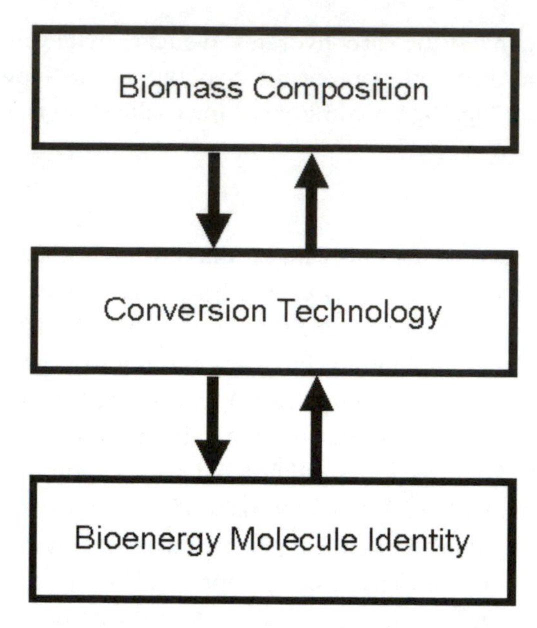

Figure 5.2 *Interactions between biomass, conversion technology and fuel molecule*

Table 5.2 *Examples of different stages of development of biofuel technology*

	First generation	Second generation
Crop	Maize (starch)	Woody biomass
Conversion	Fermentation	Thermochemical or biochemical
Fuel	Ethanol	Higher alcohols/alkanes

Biofuels that are identical in chemical properties would be best to substitute for gasoline (C_5-C_{12}), jet fuel (C_8-C_{16}) and diesel (C_{10}-C_{22}). A wide range of chemical and biochemical processes (Table 5.3) has been developed for conversion of plant biomass to fuels (Huber et al, 2006).

Ethanol has been widely produced and used as a biofuel and can be considered a first-generation biofuel product. Growth in ethanol production

Table 5.3 *Biofuel technologies and products*

Technology	Type of fuel produced
Fermentation from sugars/starch	Ethanol (butanol)
Extraction from oil containing plants	Biodiesel
Fermentation from plant cell walls	Ethanol (butanol)
Thermo chemical (Fisher-Tropsch)	Bio-oil
Chemical conversion of plant carbohydrates	Dimethylfuran

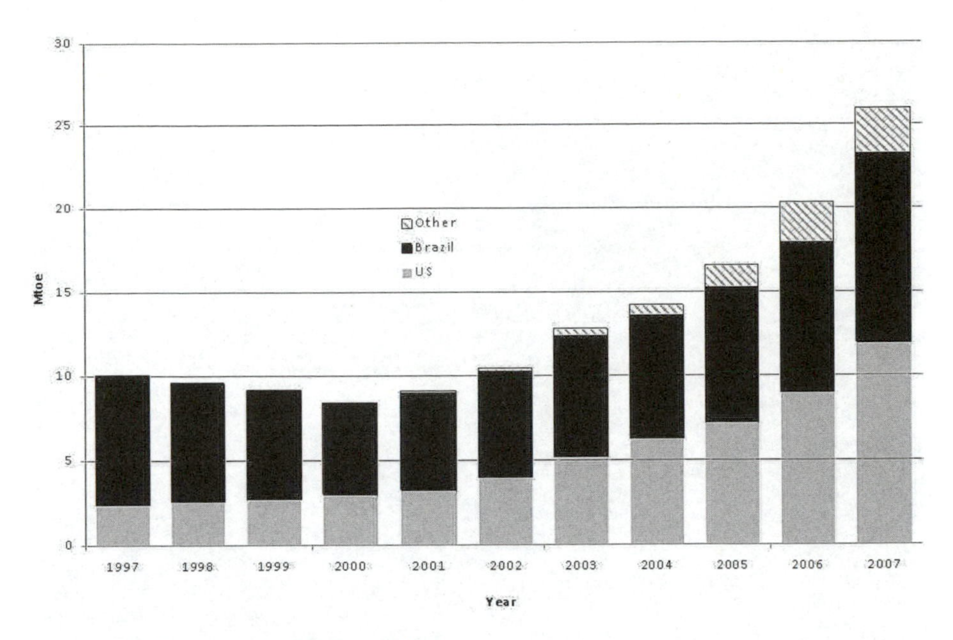

BP Statistical Review of World Energy, 2008

Figure 5.3 *Global ethanol production*

worldwide is shown in Figure 5.3. Most production is in Brazil (from sugarcane) and the US (from maize). Ethanol from sugarcane is widely available in Brazil (Figure 5.4). The current production is also all first-generation technology being based upon the conversion of sugar and starch to ethanol by fermentation.

Ethanol has long been produced by fermentation of sugars using microorganisms in brewing and wine making, and these technologies have been perfected by humans over many thousands of years. Most vehicles can operate with the addition of a small amount of ethanol (5–10 per cent) to the gasoline; vehicles can be readily produced, however, to operate on 100 per cent ethanol. Ethanol produced from cellulose rather than sugars or starch is considered a second-generation biofuel in that it is produced using a second-generation technology. The conversion of cellulose to fuel is much more technically challenging than the processing of simpler sugars and soluble polysaccharides (Lynd et al, 2002). The efficiency of this process on an industrial scale is the subject of considerable research efforts because of the abundance and low cost of cellulosic biomass. Bacteria have recently been engineered to produce more attractive fuel molecules such as 1-butanol, 2-methyl-1-butanol, 3-methyl-1-butanol and 2-phenylethanol from glucose (Atsumi et al, 2008).

Figure 5.4 *Ethanol produced from sugarcane on sale in Brazilia*

Ethanol has several disadvantages as a biofuel:

- much of the CO_2 (two-thirds) contained in plant carbohydrates is released into the atmosphere during fermentation;
- ethanol does not contain as much energy as other potential biofuels;
- ethanol is also hydroscopic and adsorbs water from the atmosphere during storage and transport.

However, the addition of ethanol may improve the environmental impact of petroleum use by reducing the emission of particulate matter and carbon monoxide. Biofuels that do not require the development of new infrastructure for distribution and storage are highly desirable. Higher alcohols such as butanol (C4) or even higher alcohols may be considered likely second-generation biofuel products, have a higher energy value and are not hydroscopic. Alkanes are still better options and may be the third-generation biofuel products. Each of these stages offers real technical advantages, but is difficult at present and requires more technical innovation. A further advantage of alkanes rather than molecules like ethanol is that alkanes are not water soluble and separate to float on the top of an aqueous production vessel. A large energy cost is associated with the separation of ethanol from water and this may be avoided if alkane-producing bacteria of fungi can be developed for biofuel production from plant biomass. However, ethanol as a first-generation biofuel product has a significant first-to-market advantage and other fuels may find it hard to displace ethanol from the market despite their technical superiority. The direct use of plant oils as biodiesel is another example of a first-generation biofuel that can often be produced by simple extraction of the oils from the plant, followed by a simple chemical conversion process to release fatty acid esters from the triglycerides. These fatty acids are then trans-esterified to form methyl or ethyl esters. Non-fuel markets may derive other molecules directly from plants. For example, fermentation may be designed to generate propanediol as a feedstock for polyester production for use in products such as carpets, and plants can be used to produce a range of chemical feedstocks that are currently sourced from oil.

Plants as sources of chemical feedstocks

Rubber is an example of a polymer that has been produced from plants. High oil prices will continue to make the traditional production of rubber from plants attractive.

Plastic production globally now exceeds 100,000 million tonnes per annum. The use of plants to replace oil for these applications is a key to reducing the cost pressures on these products resulting from rising oil prices, and ultimately making these products without consuming fossil fuels that have an associated negative impact on global climate.

Two distinct options are available for using plants instead of oil to produce plastics. Firstly, plant biomass can be harvested and processed to produce the required chemical feedstocks. The second option is to generate the chemical feedstock molecules directly in the plant following appropriate metabolic engineering.

An example of the first type (biomass conversion) is the production of PLA. Polylactic acid (PLA) is produced from corn starch by hydrolysis to glucose, fermentation of glucose to lactic acid, dimerization to produce lactide and then polymerized to produce PLA.

An example of the second type (production in the plant) is the production of PHA. Polyhydroxyalkanoate (PHA) production has been demonstrated in switchgrass (Somleva et al, 2008) and sugarcane plants. These renewable and biodegradable plastics may be produced as a co-product with fuel derived from the lignocellulosic biomass.

Other biopolymers that might be produced in plants include poly-amino acids and fibrous proteins (Van Beilen and Poirier, 2008).

The use of plants to generate multiple products is a key strategy to ensure economic viability of plant production and processing. The concept of biorefineries has emerged. A biorefinery processes plants to a range of end products that may include energy (in various forms), chemical feedstocks (e.g. plastic precursors), food ingredients and pharmaceutical compounds. The total value of the products may make the process economically and environmentally desirable even if each of the separate products is not viable. Development of plants that can be processed in this way imposed complex requirements on the plant breeder.

Plant species for bio-energy production

Many criteria may be devised to define the optimal plant species for use in biofuel production:

- high biomass accumulation;
- high harvest index;
- high fraction of biofuel in harvested biomass;
- nutrients partition to non-harvested parts;
- able to be grown on marginal lands;
- harvested material able to be stored in the field;
- high bulk density;
- high water use efficiency;
- high N use efficiency;
- potential as a weed;
- co-product potential;
- biomass composition;
- scale of potential production;
- cost of harvest;
- suitability for genetic improvement.

Most biofuels produced to date have utilized only a small number of species. Sugarcane has been used in Brazil, maize in the US, canola (and sunflower) in Europe and soybean in the US. A wide range of plant species are currently being developed or evaluated as bio-energy crops (Table 5.4). Several of the most well-known options are now introduced.

Maize

All of the cereals are rich in starch and as such are potential sources of biofuels (Henry and Kettlewell, 1996). However, cereal such as rice and wheat are

Table 5.4 *Examples of plants that have been considered as a source of biomass for biofuels*

Grasses		
Sugarcane	*Saccharum* X	Current leading industrial crop
Miscanthus	*Miscanthus* sp	Relative of sugarcane
Switchgrass	Panicum virgatum	C4 dedicated energy crop
Giant cane	*Arundo donax*	Large grass
Cereals		
Maize	*Zea mays*	Feed/food crop (major current source of ethanol)
Sorghum	*Sorghum bicolor*	Feed/food (sweet sorghum possible energy crop)
Wheat	*Triticum aestivum*	Major food crop (straw available for biofuel)
Rice	*Oryza sativa*	Major food crop (straw available for biofuel)
Barley	*Hordeum vulgare*	Major food crop (straw available for biofuel)
Trees		
Eucalypts	*Eucalyptus* sp	Pulp and timber species 700+ taxa
Poplar	*Populus* sp	Hybrids
Willows	*Salix* sp	Hybrids
Pines	*Pinus* sp	Major timber species (waste options)
Casuarina	*Casuarina* sp *Allocasuarina* sp	Limited current uses
Oil crops		
Oil palm	*Elaeis guineensis*	Food crop
Canola	*Brassica napus*	Food crop
Sunflower	*Helianthus annuus*	Food crop
Soya bean	*Glycine max*	Food crop
Olive	*Olea europea*	Food crop
Camelina	*Camelina sativa*	Non-food crop
Jatropha	*Jatropha curcas*	Non-food
Castor oil	*Ricinus communis*	Non-food
Safflower	*Carthamus titctorius*	Non-food
Jojoba	*Simmondsia chinensis*	Non-food
Diesel tree	*Copaifera langsdorfii*	Non-food
Pongamia tree	*Milletia pinnata*	Non-food
Other crops		
Sugar beet	*Beta vulgaris*	Tropical cultivars under development
Acacia	*Acacia* sp	Non-food
Cassia	*Cassia* sp	Non-food

very attractive human foods and food end uses are likely to continue to dominate for these species. Maize is also important as a human food regionally (e.g. in Africa and Meso-America), but on a global scale is less directly a human food than wheat and rice being used predominantly as a feed for domesticated animals that in turn supply food to humans. Maize was possibly originally domesticated as a source of stalk sugar (Smalley and Blake, 2003) in the same way as sugarcane (a relative of maize) and was only later selected for grain. Maize has recently been used to produce ethanol on a large scale, mainly in North America. Maize is currently a major source of first-generation biofuel globally. Maize cultivars especially suited to fuel production are being developed. First-generation biofuel varieties are developing the grain as a source of fuel and second-generation varieties are targeting improvements in the total plant biomass.

Sorghum

Sorghum is adapted to production in hot and dry environments that are not generally suited to the production of major food crops, making it an attractive option for bio-energy. However, sorghum grain has become an important animal feed in many areas, creating a potential for competition between animal feed and energy production in the utilization of sorghum as in the case of maize. Sorghum cultivars are also being specially developed for fuel use.

Sugarcane

Sugarcane (*Saccharum* X) is a major sugar and energy crop. World production is high and growing, especially in Brazil (Figure 5.5). This C4 plant

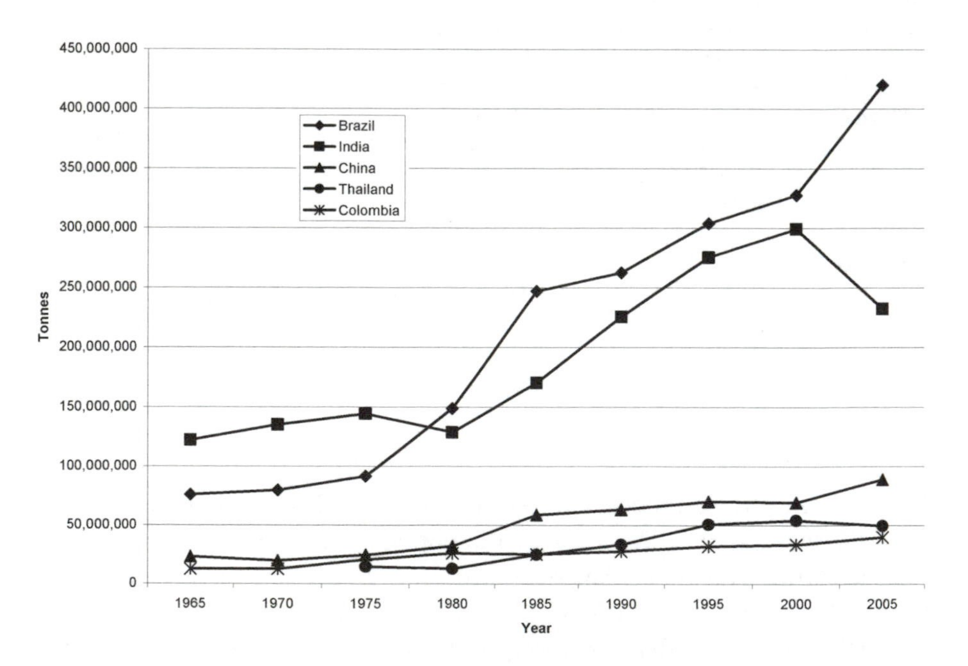

Figure 5.5 *Sugarcane production*

Sugarcane is the world's leading industrial crop. Much more than 1,000,000,000 tonnes per year are harvested worldwide.

Figure 5.6 *Sugarcane*

gives very high biomass yield in tropical environments (Figure 5.6). Sugarcane has been selected for sugar production, while sugarcane for energy – 'energycane' – does not require sugar (assuming efficient cellulosic conversion technologies are available). Recent efforts in the selection of new energycane genotypes for biofuel production as an alternative to the existing sugarcane genotypes recognizes that selection for high sugar content in sugarcane has not been entirely consistent with obtaining the maximum biomass as required for an energycane.

Ethanol production from sugarcane currently relies on processing the sugar or other co-products but not the residual fibre. When the price of sugar is high it is attractive to sell the purified sugar as sucrose rather than convert it to ethanol. Co-products such as molasses (a sugar-rich residue from sugar refining) may still be used for ethanol production in these processes. Active research programmes are targeting the use of the fibre by developing commercial-scale cellulosic conversion technologies specifically targeting sugarcane. This use competes with the widespread use of the fibre to generate electricity.

Miscanthus

Miscanthus is a close relative of sugarcane and it is being developed as a dedicated energy crop. Hybrids between sugarcane and Miscanthus have been produced as potential energy crops. One of the advantages of this species is the potential to grow the crop for up to 15 years before replanting. The ability to harvest repeatedly without the need for replanting greatly improves the performance of the system in life-cycle assessment and reduces the energy and greenhouse gas impact of fuel production from this type of crop. Miscanthus (*Miscanthus* X *giganteus)* has been shown to have much higher yields than switchgrass (more than 20 tonnes per hectare for Miscanthus compared to 10 tonnes per hectare for switchgrass) in Europe and North America (Heaton et al, 2004). This high yield makes Miscanthus a very strong candidate crop for use in biofuel production.

Switchgrass

Switchgrass (*Panicum virgatum*) is a perennial grass that has been evaluated as a bio-energy crop. Switchgrass is a C4 plant that is a native of North America, being found from Mexico to Canada. The plant is polyploid, self incompatible and highly heterozygous with great potential for further improvement by current active plant breeding. A major advantage of switchgrass is that it can be produced on marginal farm land (Schmer et al, 2008). Switchgrass is being bred for use in energy production.

Other grasses

Many other grass species are being evaluated as potential bio-energy crops in different regions. Globally there are around 10,000 grass species. Many of these have been evaluated as potential food crops or pasture species, but not as energy crops. Grasses offer advantages in wide adaptation, allowing very large-scale production. The mechanized harvesting of grasses is a well-established technology that provides an advantage relative to some other options such as trees or shrubs. Systematic evaluation of the options for energy production from grasses will require the establishment of the criteria that are required for selection of superior types. The domestication of new species for energy use provides an opportunity to use many species that have not been domesticated for food use. However, some of the same traits will be important. For example, seed retention (non-shattering) will be required to allow harvest of seed for the propagation of the crop. Traits such as seed size are more likely to conflict, large seed has been a key to food use (of the seed), while larger seed may not be associated with the maximum biomass accumulation required for energy use.

Tree crops

Wood is an attractive biomass source for bio-energy production because it has a relatively high bulk density. This makes it highly competitive with other less dense biomass sources such as plant leaves (a major part of grass biomass) because of the reduced costs of transport and handling. A key issue for trees is how long the tree should be grown before harvest. Many forest trees grown for timber (solid wood products) require 20 to 50 years before harvest and these long crop cycles make economic production difficult. More frequent harvest of trees grown at higher density is an option that has not been fully explored for many tree species. Ideally the tree can be harvested and allowed to regrow for repeated harvesting without replanting. This approach to plant cultivation is not new, but has not been optimized for most species being considered as energy crops. This depends upon the biology of the tree and its ability to grow from the base following harvest. The optimal harvest frequency could be as low as one or two years for fast growing species than can be mechanically harvested if they are not too large. The harvest frequency needs to be optimized to deliver the maximum sustainable yield of biomass per year. The costs of alternative harvesting strategies and frequencies also need to be considered, together with the impact of harvest frequency on biomass composition and resulting suitability for biofuel production.

Poplar

Hybrid poplars are potential energy crops in northern areas. The availability of the genome sequence makes this species a target for understanding the genetic control of useful traits (including fuel traits) in tree species.

Willows

Willows (*Salix* species) have been suggested as energy crops especially in the UK. They may represent a major woody biomass option in suitable environments, but have complex genetics that will make breeding challenging. Harvest every 2–3 years should be possible, making this an attractive option for biomass production in environments that suit willow species.

Eucalypts

The Eucalypts are a group of 500–1000 species originating in Australia and adapted to a wide range of climates, including many that are marginal for use in agricultural production of food crops. Eucalypts are widely planted as a forest species and have been used as pulp for paper and as solid wood for construction. They represent ideal candidates for development as bio-energy crops. The biomass yield from species such as Eucalypts may be advanced in stages. Wild material can be screened to identify genotypes that have high growth rates. The management of the production system may then be adjusted to maximize biomass yield per hectare per year. Eucalypts can regrow from a lignotuber, allowing coppicing and repeated harvests at a frequency designed to achieve the highest possible sustainable biomass yield. Plant breeding could be used to produce genotypes that perform better under these management systems. Eucalypts grown in a conventional rotation are shown in Figure 5.7 in comparison with a related species harvested annually. Eucalypts have become so widespread that they are considered a weed in many parts of the world. This may result in some concern about their widespread use as an energy crop.

Casuarina

The Casuarinaceae includes trees and shrubs that grow in areas with low soil nutrients (e.g. *Allocasuarina*) and low or variable rainfall. These plants are being evaluated in several countries because of their high nitrogen efficiency due to associations with microbes in the soil.

Oil crops

Most oilseed crops that have been developed for other applications, especially food, have also been considered as options for bio-energy crops. Soybean is an important example, with the protein component being a very important co-product with the oil. Oil-producing plants, like other grains, are probably not good options for biofuel production because of their relatively low fuel yield per hectare of land or unit of water consumed.

Camelina

Camelina sativa is a member of the Brassica family (Brassicaceae) with a seed containing around one-third polyunsaturated oil. Camelina is being developed as a non-food oilseed crop suitable for more marginal production environments. This may become a dedicated energy crop, but could also be developed as a more traditional oil crop for other end uses.

Upper panel: Eucalypt plantation in South Africa. Eucalypts, members of the Myrtaceae family of plants, are a major source of woody biomass worldwide but especially in warmer and dryer areas. These trees are harvested after many years of growth. Lower panel: a Tea Tree Plantation in Australia. Tea Tree (*Melaleuca alternifolia*), a woody plant from the Myrtaceae, has been adapted to annual harvest for the production of oil from the leaves. Mechanized and repeated harvesting of this species is possible with short growth cycles. This type of production system may suit other woody biomass crops for energy production, allowing total above-ground biomass to be utilized.

Figure 5.7 *Comparison of conventional cultivation of Eucalypts and growth of a related Melaleuca for annual harvest*

Jatropha

Jatropha curcas is a weedy shrub from the Euphorpiaceae that can be grown in marginal areas. The seeds are toxic, thus preventing their use in human food. This species has received widespread publicity and interest. The seed contains an oil that is suitable for use as a source of biodiesel. Oil crops such as Jatropha are unlikely to be competitive long term with second-generation biofuels produced from high biomass crops. Production is labour intensive, with very low yields of seeds high in oil.

Palm oil

The oil palm (originally from Africa) has been widely cultivated for food oil. Use as a fuel has been considered controversial because of the potential conflict with food use and the risk that sensitive tropical environments with high biodiversity values might be used for increased oil palm production. Again, this crop may allow production of biodiesel from the fruits and seeds, but is unlikely to be a sustainable competitor with high biomass crops delivering second-generation biofuels. A yield of around 10 tonnes per hectare of fruit may be produced, but this represents only about 3 tonnes of oil per hectare. More sustainable biofuel crops in these high rainfall environments could potentially produce more than ten times this fuel yield per hectare. The environmental footprint of oil crops is generally excessive given the competition for land and water with food and biodiversity conservation.

Castor oil

The castor bean (*Ricinus communis*) is native to the Caribbean and central America. A product, Nylon-11, is manufactured from castor oil extracted from castor beans and is used in powder coating (Ahmann and Dorgan, 2007).

Canola

Canola has been described in Chapter 2 as an important food crop. Significant quantities of Canola are now being produced for biofuel production, especially in Canada and Germany.

Pongamia

Pongamia pinnata is a leguminous tree from the Indian subcontinent and south-east Asia that is now grown widely in the tropics. The seed has 30–40 per cent oil and the trees can be grown on marginal land, suggesting that this may be suitable as a biofuel crop (Scott et al, 2008).

Algae

Both green algae and blue-green algae (cyanobacteria) can produce hydrogen – a fuel that would allow the carbon cycle to be avoided. Developing this as a commercial process will require substantial research and development. The commercial-scale production of biofuels based upon the lipids from algae is the subject of intense research effort. The biochemical pathways in these systems are now well known. More than 40,000 species have been described and many more are known to exist. The production of protein co-products may be an important contributor to the economics of algal biofuel production systems. Algae may have a use in the capture of CO_2 released during fermentation of carbohydrates from plants to produce biofuels. This could provide a valuable carbon source for the growth of algae for biofuel production. A major limitation in the advancement of this technology has been the observation that when algae are selected for higher oil content they are found to grow more slowly. High oil strains are likely to be overgrown by lower oil content strains in all but high input systems designed to exclude them. Energy can either be used to produce oil or to support growth, but not both.

Algae production from salt water (sea water) is more attractive as it avoids the issue of freshwater reserves. However, open production systems suffer from evaporation leading to salt accumulation, and closed systems designed to prevent water loss may suffer from poor light penetration.

Energy efficiency of bio-energy production

The energy efficiency of the production of different fuels from different plant sources is an important basis for evaluating the value of specific production systems. First-generation biofuel production has generally provided low-energy returns, but this is now being improved in many existing facilities. This analysis needs to consider the energy input required to grow the plant and convert it to fuel relative to the energy value remaining in the fuel. In some situations energy balance may be less important than the impact the overall process will have on greenhouse gas emissions. It may be that energy-inefficient systems that avoid greenhouse gas production are of value in reducing potential climate change. However, the cost of inefficient processes may be prohibitive. The distance that the biomass needs to be transported for processing is a key issue (Figure 5.8). The energy cost of transportation limits the distance that biomass can be moved – bulk density of the biomass is important with low-density materials requiring too much energy to transport long distances. The economies of scale of biofuel conversion are also important since the distance that biomass needs to be transported depends on the capacity of the facility or the quantity of biomass required by a conversion facility for year-round operation. The amount of biomass that can be produced close to the facility depends very much on the yield per hectare. The advantage of high-yielding crops can be considered in two different ways:

1 the distances required to source a given quantity of biomass will be reduced; or
2 for the same transport distance, the capacity of the plant can be increased to take advantage of economies of scale.

Box 5.2 *Biomass transportation*

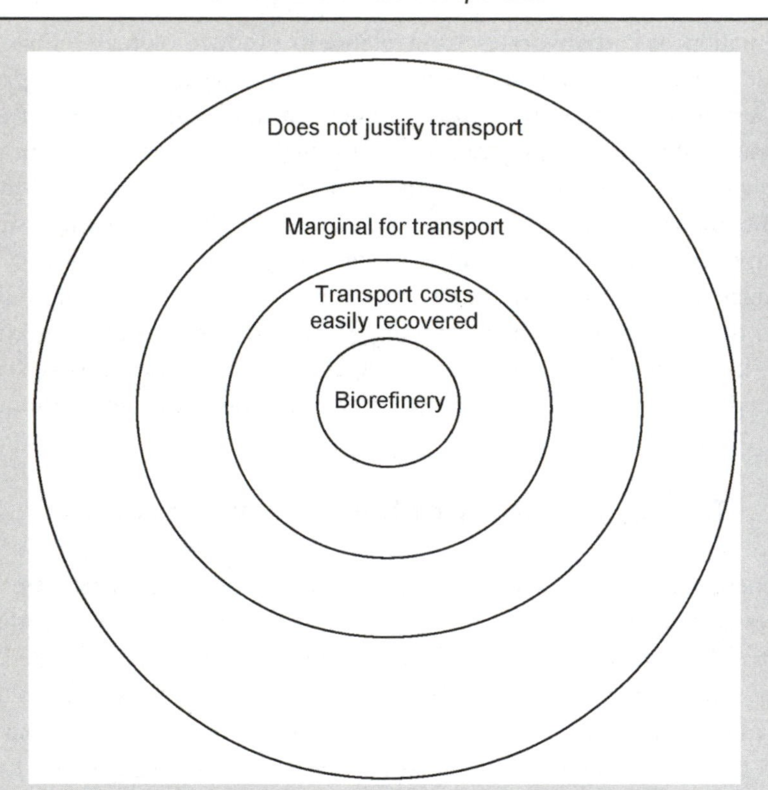

Figure 5.8 *Transportation of biomass for biofuel production*

Transport of biomass consumes energy that reduces the net energy value of the biofuel production process. Higher yielding crops allow more total volume of biomass to be produced within a distance that justifies the energy cost of the transport and allows a larger bio-refinery with greater economies of scale to be built. Crops with a greater bulk density and with a greater product yield per ton transported will also justify transportation over greater distances.

The movement of pyrolysis oil in pipelines to central facilities for further processing is an example of the type of system being devised to cope with these constraints.

The energy costs associated with the separation of fuel molecules (e.g. the separation of ethanol and water) may be a very important aspect of the energy efficiency of the process. Micro-organisms used for fermentation produce ethanol only until the concentration of ethanol becomes toxic to the organism. Distillation of ethanol from the dilute solutions that can be produced by fermentation may require significant energy. Use of renewable energy such as solar energy as a source of heat for these processes is being explored and alternative separation technologies are also an active area of current research.

Thermochemical conversion technologies are conducted at high temperatures and this may require excessive energy inputs.

Carbon balance of bio-energy production

A key measure of the value of biofuel production in reducing greenhouse gases is the assessment of the carbon balance of the whole life cycle of production of the plant and the fuel. Each plant system will have different efficiencies.

An important innovation that may improve the carbon balance associated with biofuel production from biomass would be the capture and sequestration of CO_2 produced during biofuel production (fermentation).

Environmental impact of bio-energy production

Biofuel production may have a significant environmental impact. A narrow analysis considers the energy and greenhouse gas impacts. However, a wide range of environmental factors need to be considered when comparing biofuel production systems (Scharlemann and Laurance, 2008). The impact on land that contributes to biodiversity conservation and contributes positive environmental benefits is a key factor and will be considered in Chapter 9. Some biofuel production systems contribute to undesirable emissions such as nitrous oxide. Water consumption and impact on water in the environment is a key impact in some regions. The issues are essentially the same as those facing all plant production for food or non-food uses.

Future demand for plants as a resource for biofuels

Continuing developments in technologies for biomass to biofuel conversion will impact on the demand for different types of biomass. Changing and more ambitious target fuel molecules may also become more realistic for biofuel: this will define the targets for biomass crop development in terms of biomass composition. The total relative environmental impact of different biofuel crops will need to be evaluated. These constraints will ultimately determine the demand (quantity and quality) for biofuel crops that might emerge. The iterative interaction between fuel (molecule target), conversion technology, biomass composition and biomass availability is likely to continue and generate several generations of biofuel technology.

The potential of plants to meet demand for biofuels

Total global biomass production greatly exceeds that required to replace all human energy needs. However, competition for food production and the need to conserve plants for environmental reasons limits the portion that could realistically be used for energy. Transportation uses only a part of total energy consumption; replacement of transportation fuels with fuels produced from plant biomass would be possible, but how we do it would determine the impact on food production and the environment.

Global energy production in 2006 has been estimated at 11,741 million tonnes of oil equivalence (International Energy Agency, 2008). Consumption of energy was 8084 million tonnes of oil equivalent, with 43 per cent of this (around 3500 million tonnes) being oil. This can be compared with the current world production of around 2000 million tonnes of cereals for human food (FAO, 2007). Human energy needs represent a requirement for biomass that is similar in scale to that required for food. The important difference is that food uses require plants with a very specific composition, while ligno-cellulosic energy production is possible from a much larger proportion of available plant biomass.

Development of improved plants for energy production

Many plants being used or considered as biofuel sources have been introduced above. However, it is likely that many other species that are better suited to energy production could be identified by careful screening of the large number of plant species. Systematic evaluation of plant species against carefully defined criteria is necessarily a time-consuming process since growth rates need to be assessed in different environments and over the entire growth cycle of the plant. The development of more suitable plants for the production of biofuels will be built on our growing understanding of these organisms at the genome level (Rubin, 2008). The technologies for the selection of superior genotypes are now possible using DNA analysis methods that continue to improve (introduced in Chapter 2). The costs of analysis have been greatly reduced and the amounts of data that can be collected have increased dramatically. The extraction of DNA from plants and the analysis of the DNA can now be applied to large plant populations using highly-automated laboratory analyses. These methods have been developed with the improvement of crops for food uses as a primary target. However, these advances will also allow the rapid development of energy crop cultivars. Accelerated domestication with current molecular technologies should allow the domestication of energy crops in a few short generations much more rapidly than was achieved by simple human selection in the domestication of most food crops over the last 10,000 years or so. This option will be discussed in more detail in Chapter 10.

The application of molecular genetics and genomics to plants is continuing to accelerate with technological developments (Henry, 2009a, b). Dramatic recent improvements in DNA sequencing technology have greatly increased our ability to analyse the genomes of plants. The development of high-throughput genotyping technologies (Henry, 2008) and improved targeted mutagenesis methods (Cross et al, 2008) increases our chances of developing the required genotypes for biofuel production.

The efficiency of biofuel production from plants may be improved by the expression of enzymes that degrade the plants' carbohydrates within the plant. In first-generation biofuel crops, the expression of amylases to degrade starch is an example of this approach. Second-generation crops may be engineered to express enzymes that can degrade the cell walls when required. Expression of a cellulase (from poplar) in a rapid growing tropical legume, Senagon (*Paraserianthes falcataria*), has been shown to result in increased growth rates (Hartati et al, 2008). This species is an example of the potential to develop novel biofuel crops that could be used to deliver highly efficient biomass production without the need for nitrogen inputs and with accelerated potential for processing to fuel. A key to these strategies may be the development of appropriate mechanisms to control the timing or location of enzyme expression in the plant.

The production of bio-energy from plants is an area in which there are potentially many different combinations of plant species and conversion technology. Significant scientific advances will be required to realize much

Box 5.3 *Nanotechnology provides greatly improved tools for analysis of plant genes*

New technologies continue to expand our ability to analyse biological systems. Nanotechnology (processes that happen on a very small molecular or atomic scale) has recently allowed the development of analytical platforms for the large-scale analysis of DNA sequences. This is currently revolutionizing biology. The volumes of data that are being collected are challenging modern computing technology. DNA sequencing with a single instrument has moved rapidly from the daily collection of Kilobases (1990s) of DNA sequence (one base = one letter of the genetic code) to Megabases (early 2000s) to Gigabases (2008). Current DNA sequencing instruments can generate terabytes of raw data in a day and petabytes of data need to be stored. This trend is likely to continue (Doctorow, 2008). These advances are providing new understandings of biological systems and processes (Gravely, 2008). The technology is providing new insights into the way plants develop and respond to their environment (Lister et al, 2008).

of this potential. For this reason it is essential that we pursue all the main options in the hope of finding an effective solution. Exactly where the technical advances will come cannot be predicted – that is the nature of innovation. We cannot afford the luxury of picking winners; we must follow up on all serious options.

6

Competition between Food and Fuel Production

Biofuel output 'crime against humanity'.

(Headline, *Gulf News*, Dubai, 15 April 2008)

The emergence of biofuel production from agricultural biomass has raised concerns that biofuel production will compete with food production and result in a reduction in food supply and higher food prices. The increases in food prices in 2007–08, coinciding with the increased production of biofuels, exacerbated this perception. This increase in food prices was associated with a tightening of supply, but it was driven by many factors including increases in demand with some hoarding of food, short-term seasonal production problems and probably also economic factors such as changes in exchange rates. Oil prices also peaked in this period, increasing the costs of food production and distribution. Defining the contribution of biofuels is difficult. However, it is likely that the new end use of biofuels may result in a higher value being placed on agricultural commodities generally and this may have a long-term impact. The magnitude of these effects remains controversial with estimates of the level of impact of biofuel production on food prices varying widely. Food prices declined generally in 2008–09 together with oil prices.

Impact of biofuel production on food prices

The analysis of the influence of biofuel production of food prices is complex and difficult to define even with the benefit of hindsight: prediction of future impacts is very difficult. Recent reports have attributed widely differing components of food price rises to biofuel production, IFPRI estimating 25–30 per cent while FAO estimated 10–15 per cent (McClung, 2008). These differences can be explained by differing assumptions in the analyses.

Agricultural commodity prices have been linked by the extent to which they can substitute for one another. For example, a shortage of wheat may result in increased wheat prices, but this may also push up the price of other grains because demand for them will grow to substitute for wheat. This interdependence of prices is especially relevant to the use of grains in animal feed where grains can be substituted depending on price to produce a feed with the required nutritional composition. The price of energy commodities has also been linked because of their potential substitution. The more recent production of large quantities of fuels such as ethanol from agricultural commodities has introduced the prospect of food and energy prices becoming interdependent.

Recent trends in sugar, maize and oil prices illustrate the challenge in predicting the future relationships between energy and food prices. The increased production of ethanol from sugarcane resulted in predictions of a strong link between sugar and oil prices a year or so ago. Sugar prices have remained low, however, despite spiking oil prices, largely because of increased sugarcane production in Brazil, and they have recently risen while oil prices were low. A link between maize and oil prices seemed to emerge in early 2008 with both commodities rising strongly together. However, this correlation in price is not likely to be strictly maintained.

Biofuel production may result in reductions in food costs if biofuels can be produced at a lower cost than fuels from oil. Lower fuel prices may reduce the cost of agricultural production and food distribution. This is most likely to outweigh any increase in food costs associated with competition with food if appropriate dedicated fuel crops or dual purpose crops are developed (see Chapters 5, 10 and 12).

Land availability for crops

Estimates of the area available for the cultivation of energy crops vary widely. Marginal lands not suited to conventional agriculture are considered an important option for bio-energy production. Estimates of the area of such land that could be used for energy production are from 100 million to 1 billion hectares (Worldwatch Institute, 2006). Agricultural land use changes in response to market demands. Many areas that were previously used for cropping are now underutilized because they have not been competitive with other regions in a highly competitive global agricultural commodity market. As prices improve more of this land could be introduced back into production. For example, examination of maize production in the US in recent years shows a shift in production to more intensive and productive areas, with a decline in production in some traditional regions. This indicates that if demand was high enough, maize product could be expanded and re-established on significant areas of land in the US that are not currently cultivated.

Many areas of extensive agricultural production such as those used for grazing of animals may be suitable for cropping to satisfy higher demand,

Table 6.1 *Examples of arable land estimates*

	Total area	Potential arable land	% in use
	km^2	1000s ha	
Asia & Pacific	28,682	777,935	61.4
Europe	6806	384,220	55.6
North Africa & Near East	11,545	49,632	144.2
North America	19,295	479,632	48.6
North Asia East of Urals	20,759	297,746	59.0
South & Central America	20,541	1,028,473	13.9
Sub-Saharan Africa	24,238	1,109,851	14.2

Source: FAO, 2008b

and may also be suitable for energy crops even if they are not ideal for food crops. Land suitable for cropping is often called arable land, but just how do we define arable land and how much do we have to use for all types of cropping? While many data (Table 6.1) can be found on areas of arable land in different regions of the world (FAO, 2008b) the definition of arable land probably needs to be redefined, especially when we consider the production of crops for biofuels on land that is of a lesser suitability than that currently used for food crops. Many countries have more than 100 per cent of designated arable land under cultivation by using irrigation. For example, it is estimated that 144 per cent of arable land in North Africa and the near East is under cultivation. Egypt and Saudi Arabia have significant agriculture but would have almost no arable land without irrigation. In many other cases data suggest that much of the arable land is not currently being used for agriculture. However, much if not most of the arable land that is not under cultivation is not really available due to the presence of forests or reserves. This makes analysis of the real situation with food security and the impact of diversion of land to energy production very difficult to establish. New technologies also have the potential to create more arable land. Finding ways to deal with nutrient deficiencies in soils or to cope with hostile (toxic) soils may make previously unusable land very productive.

Brazil

Brazil is a major food exporting country with potential for continuing major expansion of production (Figure 6.1).

Brazil is a large country with a very large agricultural production and potential for expansion, and it includes diverse regions and environments.

Box 6.1 *Arable land in Western Australia*

Areas of the south-western corner of Australia are used for the production of cereal crops, mainly wheat. This area has a Mediterranean climate (wet winters and hot dry summers) and generally sandy soils of very low fertility. Many areas that had been considered unsuitable for cropping have become productive by growing a legume (lupins) that greatly increases the yield of wheat that can be obtained from these areas (wheat yields three times higher have been achieved by using the legume in the rotation). This has made land arable that was formerly not, by increasing nitrogen and providing crop residues that improve soil carbon and key nutrient levels in the soil. Crop rotation is an essential part of sustainable agriculture in many regions.

My first exposure to the cereal production system in Western Australia was at a meeting at a plant breeding research station in Wongan Hills. The sandy soils and relatively low crop biomass led me to comment to the scientists at the meeting that the poor crop may have been due to the poor soils. My experiences had been with cereal crops in eastern Australia, especially the black clays of the Darling Downs in Queensland that produced much higher yields. The extent of my faux pas became evident when they later told me that not only was this equal to their best soil, but that crop was also probably among the highest yielding crops in the region. This example illustrates the complexity of defining arable land. The definition is almost always a relative one and may differ greatly between regions. New technology or the option of new crops can make previously non-arable land arable. The amount of land available for food and energy use can really only be established on a local basis after consideration of all of the cropping options that might work locally. All of this makes estimation of the total food and bio-energy production potential of the Earth very difficult to estimate objectively. New tools and processes are required to answer these important global questions.

The Amazon rainforests are not a likely contributor of significant land for agriculture as the biodiversity values are very high and the costs of land clearing excessive given the short life of production on the resulting areas. The major area that is likely to be used for expansion of sugarcane production is the Cerrado, an extensive area of savannah in the south of Brazil. Significant areas of land with adequate rainfall are available for cropping in this region without the need to clear native vegetation. Many of these are degraded pastures that have already been cleared of the original native

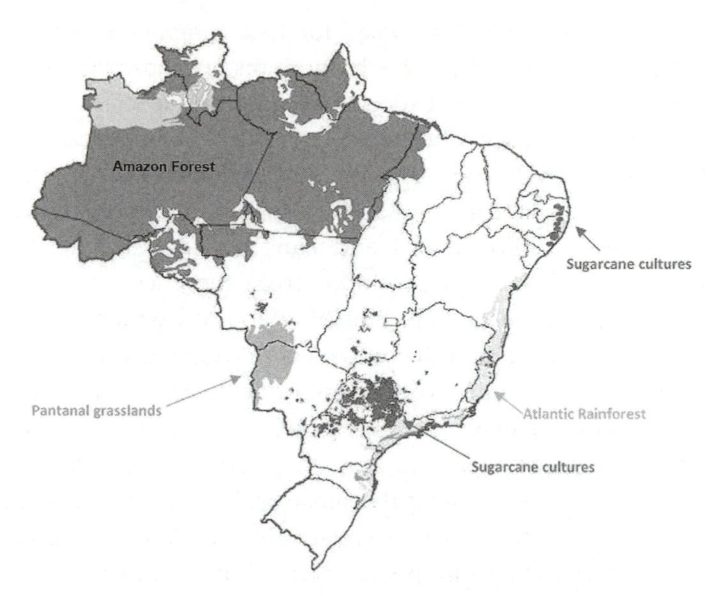

Source: Goldemberg, J. (2008) *The Brazilian biofuels industry*. Biotechnology for Biofuels 2008, 1:6doi:10.1186/1754-6834-1-6 at (http://www.biotechnologyforbiofuels.com/content/1/1/6).

Figure 6.1 *Brazil – land use*

vegetation and used for extensive animal production. However, this expansion may still put pressure on biodiversity with many plant species endemic to this region.

Mexico

Mexico has significant areas that have not been used for food crops but may suit energy crop species. The rainfall in many of these areas may be sufficient but also highly variable. These areas that are marginal for the production of food crops may be available and suitable for perennial energy crops.

Australia

Significant areas that have had the native vegetation removed for grazing could be suitable for energy crop production. Rainfall is highly variable in these areas requiring the careful selection of species.

Multi-purpose or specific purpose crops

The competition between food and fuel can be defined in relation to impact on total crop production, the type of crops produced and the influence on the market (supply levels and price). Plants that contain sugar and starch (non-structural carbohydrate) are used for first-generation biofuel production. These plants are also digestible by humans and other animals, making them generally favoured as food species. In contrast, species that are rich in cell wall carbohydrates such as cellulose (structural carbohydrate) are essentially not digestible by humans and are not major components of human diets. Ruminant animals (e.g. cows and sheep) use micro-organisms in their gut to digest these carbohydrates and can therefore use them as food. Second-generation biofuels are produced from these less attractive human food sources. Second-generation biofuel will not compete directly with major human food crops, but will compete with ruminants for biomass. This creates a potential for indirect competition between these animal-based foods and biofuel production. Cellulosic biomass can often be produced in more marginal areas not suitable to food crop production. Extensive pasture areas used for grazing are among the most likely to be adopted for biofuel production. However, these areas are relatively abundant and significant co-existence of biofuel and pasture production will be possible.

The potential for multipurpose crops needs to be explored for crops with significant levels of both structural and non-structural carbohydrate content. For example, grain crops could allow harvest of the seed (grain) as a food (rich in non-structural carbohydrates of high nutritional value to humans and animals) and the remaining parts of the plant as biomass (straw containing most of the structural carbohydrates) for biofuel production. Breeding of grass cultivars to suit this dual purpose use may optimize this option. However, the major gains in food production in the last 50 years have been achieved by increasing the harvest index of cereals (Chapter 2). The harvest index is the proportion of the plant biomass harvested as the edible part. The relatively high value of food relative to other biomass suggests that it would not be attractive to reduce the harvest index in dual purpose crops. The secondary or by-product role of the biomass in these crops indicates that improvements in the composition of the residual biomass could be an important innovation. Crop residues remaining in the field after harvest are important for soil properties and the sustainability of the farming system – removal of these crop residues may be highly undesirable. Dedicated energy crops with high biomass potential, especially in areas not suited to efficient food crop production, will remain attractive priorities.

Life cycle assessment of cropping systems

The environmental desirability of both food and energy crops needs to be examined by analysis of the entire production system. Life cycle assessment

(LCA) examines the environmental impact and economics of the entire production system. Defining the limits of such systems appropriately is a key to relevant analysis. For example, the environmental and economic costs of producing nitrogen fertilizer used in crop production need to be considered. However, the environmental costs of manufacture of the clothing worn by the workers in the nitrogen plant should be outside the analysis. These costs to the environment would probably be incurred even if the workers did not make fertilizer.

Direct and indirect land use changes

Direct land use competition may be avoided, but indirect competition may apply. For example, use of land that has not previously been cultivated may represent indirect competition. Displacement of animal production, or increased reliance of animal production on grain or other feeds rather than forage, represents another form of indirect competition.

Future competitive scenarios

Cline (2007) has analysed the situation comparing projections for 2085 with 2005. This illustrates the factors that need to be considered in predicting how we might meet future demands for food and energy globally. These calculations are unlikely to predict the future. Factors that we cannot predict at this time are likely to alter the absolute outcome of different scenarios. The projected increase in demand or supply may not reflect what eventually happens over these extended periods of time. However, the comparison of different options allows an evaluation of the relative value of alternative directions and provides a basis for evaluating different policies or research targets.

Growth in food demand

Population is predicted to grow to 10.5–14.7 billion by 2085. This represents a 1.63–2.28-fold increase in population. Per capita income is growing and at a rate which would give a 1.63-fold increase in demand for food per capita by 2085.

Taking these two factors together gives a 2.66–3.72-fold ($1.63 \times 1.63 - 2.28 \times 1.63$) increase in demand for food by 2085.

Growth in supply of food

Diversion of land from food production to energy supply is estimated to alter supply of food by a factor of 0.7 by 2085.

Increased yields derived from the application of improved technology are estimated to give a 3.48-fold increase in food production by 2085.

These two factors give an overall growth in supply of 2.44-fold (0.7 × 3.48).

The growth in demand (2.66–3.72) exceeds the growth in supply in these predictions.

These scenarios allow little room to cope with further losses in agricultural productivity due to global warming.

Several aspects of this scenario are worth analysis as we consider the approaches we should take in trying to shape a workable future. The potential of technology to deliver increased agricultural productivity may be considered optimistic if we consider recent trends. The rate of growth in agricultural productivity has slowed in recent years as the influence of the Green Revolution has worked through and better management practices have been more widely adopted in agriculture. This trend may continue and increases may slow further in the future.

Cline (2007) also points out that these projections may not fully account for the impact of changes in diet towards the consumption of more meat. This could add further significant increases in demand.

This approach allows the impact of different levels of diversion of land for energy production to be considered. Table 6.2 details a range of future scenarios based upon simple arithmetic adjustments to the numbers generated by Cline (2007). This approach (assuming all land is equal) is likely to overestimate the impact of diversion of land to energy production if less favourable land was used. However, the diversion of around 30 per cent of current agricultural production (the intermediate option) would put great pressure on food supplies.

The intermediate scenario clearly indicates that significant areas of land not currently being used for cropping would also need to be used for energy cropping to supplement the 30 per cent diverted from food production (the impact of this on biodiversity will be considered in Chapter 9), and significant improvements in the technology are required to deliver the promise of renewable fuels from plants. Advances in technology could make this scenario very realistic. The improvements in technology required to make this scenario work will probably be delivered if we are able to perfect ligno-cellulosic conversion (as described in Chapter 5).

With current first-generation technology we currently produce only a very small amount of biofuel – less than 1 per cent of transport fuels – from

Table 6.2 *Impact of different levels of biofuel production on food supply (assuming no expansion in agricultural land)*

Population	Diversion of land to Biofuels	Food supply	Food demand	Gap
Low	None (0%)	3.48	2.66	+ 0.82
(10.5 billion)	Intermediate (30%)	2.44	2.66	–0.22
	High (50%)	1.74	2.66	–0.92
High	None (0%)	3.48	3.72	–0.22
(14.7 billion)	Intermediate (30%)	2.44	3.72	–1.28
	High (50%)	1.74	3.72	–1.98

All values are times 2005 levels

a very small area of land – less than 1 per cent of arable land. Diversion of all (100 per cent) of the current biomass produced in agriculture would probably be required to replace all fuels used for transportation with current technologies. However, new technologies will dramatically change this and should allow production of all transport fuel from dedicated second-generation crops and technologies.

The above analysis is unlikely to predict the future, but provides a basis for comparison of the impact of different policies in relation to agriculture for food or energy production.

We also need to consider the impact these future scenarios might have on global biodiversity. Chapter 9 will explore the competition between food and energy production further in the context of the need for the conservation of biodiversity. We will now turn to consideration of the importance of plants to the environment and biodiversity (Chapter 7) and the likely impact of climate change (Chapter 8) on these roles of plants.

7

Plants, Biodiversity and the Environment

Suddenly as rare things will, it vanished.

Robert Browning (as quoted by Leigh et al, 1984)

Plants are fundamental to life on Earth. They play a key role in determining composition of the atmosphere and are the base of the food chain supporting other higher organisms. We have explored the key role of plants in food and fibre production and their growing potential for direct use for energy supply. Plant diversity provides food and habitat for a great diversity of insects, fungi and animals. Conserving plant biodiversity is central to conserving biological diversity on earth. However, the expansion of human populations has resulted in local and global reductions in biodiversity. Species extinction on a large scale has followed human population growth. Many more species are now endangered.

Plant diversity can be considered at many levels – most often we think of it at the level of the species. However, diversity at both higher and lower levels is important. At the higher level we need to consider questions such as how different are the species we have and how are they related to one another. At the lower level we need to consider how much variation exists within the species. The extreme example in plants can be that no diversity exists within the species and all individuals are clones (vegetatively propagated). In contrast, many plant species include highly divergent individuals.

Species diversity

The number of species of plants is not well defined. In some areas new species are still being discovered frequently, yet many areas remain poorly explored for plant diversity. Discovery rates sometimes reflect the amount of analysis effort. For example, recent growth in commodity markets has resulted in increased mining development in Western Australia and an increase in the rate of discovery of new species as environmental impact assessments are conducted on potential mine sites.

Box 7.1 *Case study – Queensland, new species*

A report of the Queensland Herbarium for 2006–07 describes 55 new species and three new genera (Queensland Herbarium, 2008). The total flora of the state now includes more than 8300 vascular plant species. The report also lists 52 new records for the state. These are native species not previously recorded for Queensland.

The number of new taxa reflects the level of activity among taxonomists. The list also includes new names that result from reclassification, resulting in a total of 305 name changes, new reports and new species.

This analysis demonstrates that the number of species in a region is not a good measure of the rate of loss of species or extinction since many factors contribute to changes in species numbers over time. These include:

- naturalization of species from other regions (growth in numbers of weed species);
- identification of an extended range from previously known species,
- discovery and naming of new species; and
- reclassification to create multiple species from a single species (splitting) or a combination of many species to form one (lumping).

None of these factors increase or decrease biodiversity, but they do improve our knowledge and understanding of biodiversity.

New weeds

The movement of people distributes plants around the world, with species being continually introduced into new areas and often becoming established as weeds.

Newly recorded species known from neighbouring regions are often found to have a wider distribution and are reported as new records for the flora of the region of interest. In the Queensland example above, 52 native species were recorded for the first time during the year. These species had previously been known from outside the state but are now recognized as native to Queensland. In addition, five non-native species were recorded during the year as completely new naturalizations because they had not previously been recorded in the state. However, a further 13 non-native species were also recorded as naturalized from species that have previously been known from cultivation in the state or that had been recorded as doubtfully naturalized. A further 16 non-native species joined the list of doubtfully naturalized species during the year.

New species
New species continue to be discovered, even in well-studied locations. In the Queensland example, three new genera and 55 new species were described. These are species new to science and were formally described and published.

New species from old
New species may be defined by taxonomic revisions recognizing more than one species within a single earlier species.

New names
Species can be renamed. This results from a reconsideration of relationships between species or from application of the rules of botanical nomenclature to determine that a previously used name is not correct and should be changed.

The current global status of plant diversity can be measured by the numbers of species listed as endangered. The IUCN Red List of threatened species includes species defined as threatened internationally. More than 8000 species, or 3 per cent of described plant species, are listed as threatened in 2007.

IUCN Red List

The plants that are endangered or threatened are listed in the IUCN Red List. This listing is published by the International Union for Conservation of Nature and Natural Resources and is available online at www.iucnredlist.org/. The list includes more than 12,000 species of plants. Classifications include extinct, extinct in the wild, critically endangered, endangered, vulnerable and data deficient. The list is far from complete and lists more species from developed countries (e.g. US) than from other less well-documented locations. The list has only been available on-line since 2000 when the lists for plants animals and other groups were combined and more details included. This makes updating easier and improves access to this important resource.

Box 7.2 *Coastal Fontainea – an example of a critically endangered plant species*

Coastal Fontainea (*Fontainea oraria*) is a plant from a littoral rainforest of eastern Australia and is known from only a single site. The ten adult plants are found within a 6000m^2 strip of private land that has been cleared of native vegetation but where the littoral rainforest is regrowing. Rossetto et al (2001) used DNA analysis to investigate the relationships between individuals in this population and the relationships to other species.

The DNA analysis showed that most of the seedlings found at the site were derived from a single adult tree, suggesting the need to actively encourage the reproduction of the other individuals in the population to ensure genetic diversity was retained in subsequent generations. The situation of this species is made more complex because individuals are of a single sex. The species may be threatened by a limited number of female or male trees.

The analysis indicated another population of the related species *Fontinea australis* was genetically unique and in need of specific conservation.

This illustrates an approach that is useful for the analysis of any rare or endangered population.

Classification of rare and endangered plant species

The definition of rare and endangered plant species is a relatively subjective but important task in defining targets for plant conservation efforts. The following types of categories may be defined (Environment Protection and Biodiversity and Conservation Act, 1999, Australia) (Henry, 2005a):

- **Extinct** – no reasonable doubt that the last member of the species has died.
- **Extinct in the wild** – species exists only in cultivation.
- **Critically endangered** – extremely high risk of extinction in the immediate future.
- **Endangered** – very high risk of extinction in the medium term.
- **Vulnerable** – high risk of extinction in the medium term.
- **Conservation dependent** – species is dependent on a specific conservation programme without which it could become vulnerable, endangered or critically endangered within five years.

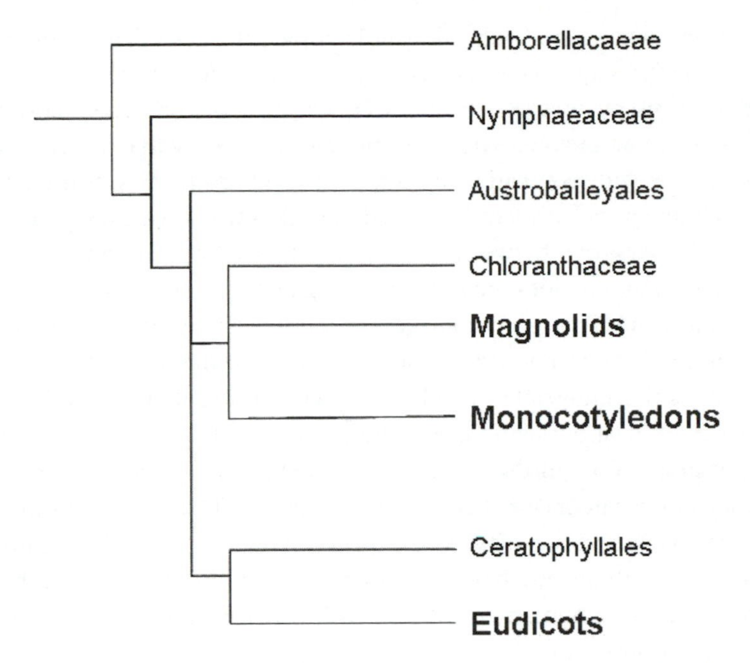

Figure 7.1 *Relationships between different groups of flowering plants*

Higher level diversity

Diversity at higher levels above the species is also important. For example, it may be more important to conserve ten species that are from divergent families (a commonly used taxonomic grouping of related plants) than ten species, all of which are members of the same family. DNA analysis is providing improved understanding of plant relationships at these higher levels (Angiosperm Phylogeny Group, 2003, Figure 7.1).

DNA analysis has confirmed most of the relationships that were originally deduced based upon the appearance (morphology) of the plant, especially the flower in the case of flowering plants. However, important refinements of our understanding of relationship have been provided by DNA analysis, resulting in significant changes in the position of some groups in the 'family tree' of plants. More details of the families of seed plants are provided in Chapter 11.

Diversity within species

Species of plants contrast in their diversity. Some vegetatively propagated species are all genetically identical (clones), but most species reproduce sexually and vary to different extents. Morphological variation is apparent within many species, while others that show little apparent variation are highly divergent at the DNA sequence level. Diversity within species provides a reservoir of genetic diversity that can be selected by the environment, allowing the species to adapt and evolve. Many species show diversity that is structured on a regional basis with populations at different locations having distinct characteristics. Other species display all of their variation within each population and little if any divergence between geographically separated populations. Plant population genetics defines these features of genetic variation and provides the understanding necessary to manage the conservation of genetic diversity within species. Knowledge that populations from different locations are distinct suggests that movement of plants between populations is undesirable if the genetics of local populations is to be protected. This will have the practical implication of requiring seeds for plantings to be sourced from local populations rather than elsewhere. Rare species may be genetically uniform or may retain substantial diversity. Some species have little diversity because they have been through genetic bottlenecks when the population has crashed due to natural events or human activities. DNA analysis allows determination of diversity in populations of rare or common plant species. Conservation of diversity within species with large populations is also important.

The level of information required to effectively manage conservation of diversity within plant species is significant. An examination of published scientific journal manuscripts (e.g. papers in Conservation Genetics) reporting on the genetic structure of plant populations and the implications for conservation demonstrates the amount of effort required to define the needs of a single species or population. These papers usually represent a substantial amount of field and laboratory research and scholarship involving commitment of significant time (sometimes many years of work) and research funds. However, they often report on a single species and sometimes only on a single population of a single species. With more than 300,000 species of higher plants, many distributed in large numbers of populations, the challenge of obtaining the knowledge to adequately conserve all plant diversity is huge. Efficient tools are required for determination of genetic diversity within and between populations (population genetics) and within and between different species (phylogenetic analysis). Knowledge of genetic selection and adaptation is also important in understanding the evolutionary processes in progress, and especially in being able to predict the likely impact of factors such as climate change on population genetics and the likelihood of long-term survival of the population or species. New technologies for DNA analysis and the more extensive use of DNA banks may facilitate the more time- and cost-effective analysis of plant population structures required to guide effective conservation efforts. However, this is just the first step towards understanding plant diversity.

Understanding plant diversity

We need to understand what we have if we are to effectively conserve it. The conservation of plant diversity requires an understanding of genetics and the evolutionary and environmental factors influencing the diversity of species and plant communities (Henry, 2006). Evolutionary processes continue the processes of speciation (the divergence or appearance of new species) and extinction (loss of species), but are increasingly impacted dramatically by human activity. The risks of extinction in small populations may be greater than has previously been estimated (Melbourne and Hastings, 2008).

Understanding the reproductive biology of the species can be a very important step towards understanding the diversity with the species, how that is distributed in the population and the likely threats to conservation of diversity in the species. For example, plants can be self-pollinating (pollination only within the flower or between flowers on the same plant) or exclusively out-crossing (only pollinating another individual), or anywhere in between (some combination of these possibilities). Knowledge of vectors for pollen such as insects, birds or animals may provide important insights into the distances over which pollination might be possible and the consequent extent of gene mixing over distance in the population. The mechanism of seed dispersal is also a key contributor to population structure, defining the extent to which populations are genetically distinct.

Box 7.3 *What is a species?*

In plants as in all other organisms this is not always an easy question to answer. We usually consider a population of inter-breeding individuals to be a species. This does not prevent the distinction of genetically different groups within a species. Species are defined by taxonomists. Traditionally this has been based upon morphology (the appearance) of the plant, with a special emphasis on the flower as a distinguishing feature in the flowering plants. Increasingly we are using the evidence that comes from analysis of the DNA. The sequence of genes (genetic code) provides evidence of evolutionary relationships.

The Consortium for the Barcode of Life (CBOL, www.barcoding.si.edu) is an international collaboration that is using DNA sequencing of agreed genes to identify species of all types of organisms. Application of this barcoding approach to plants has proved more difficult than it was for animals. A combination of the DNA sequence of more than one gene may be necessary to identify all plants to the species level. Sometimes this produces surprising or unexpected results. Species that we thought to be unrelated on the basis of appearance turn out to be close relatives. In other cases, plants that were classified as being related or even the same species are shown to be very different. These are the exception, with most DNA analysis supporting the classifications devised by taxonomists based upon morphological observations. A recent example is our study of two 'species' of spotted gum (a type of Eucalypt) – *Corymbia variegata and Corymbia henryi* – that grow in the same areas on the central east coast of Australia (Ochieng et al, 2008). Genetic analysis (DNA analysis) shows that these two 'species' are one inter-breeding population. The differences in size of the leaves and fruits that were used to define the two species may be explained by genetic variation within the population in only one or a few genes.

The factors threatening the species or community also need to be well understood so that they can be mitigated, or alternative options for conservation (e.g. *ex situ* conservation) can be implemented.

Defining threats to plant diversity

Factors causing loss of diversity or threatening extinction may be associated with long-term evolutionary processes. However, most threats of loss of diversity in the short term are due to human activities. Loss of habitat by human displacement is a dominant cause of extinction. Other causes include:

- introduction of new pests and diseases;
- grazing by domesticated animals;
- environmental pollutants;
- competition from weeds;
- altered water status of soil (drainage or construction of water storages);
- removal of canopy to encourage pasture growth for grazing;
- genetic impact of domesticated relatives.

The impact of successive colonizations of the Hawaiian Islands by humans provides a useful example of the types of impact we have had and are having on plant biodiversity.

The islands of Hawaii are unique in having a reasonable land area (10,500km^2) that is a long way from any other significant land area. These volcanic islands now have around 1000 species of native plants (90 per cent of which are endemic (found only in Hawaii)) competing with around 800 species (not including plants cultivated in gardens) of exotic naturalized species that have been introduced since European contact, starting with the visit of James Cook in 1778 (Sohmer and Gustafson, 1987). Since that time at least 10 per cent of the plant species have become extinct and about 30 per cent of the remaining flora is endangered or threatened.

This was not the first human impact upon the flora of these islands. At the time that Cook arrived, Hawaii had a population estimated at between 300,000 and 500,000. Polynesians arrived 1000 to 1500 years ago, bringing at least 27 species of plants mostly with utility. Plants introduced by the Polynesians included at least 12 food species. For example, the coconut (*Cocus nucifera*) was introduced to Hawaii by Polynesians. This human invasion probably also resulted in a wave of plant species extinctions.

Weeds and biodiversity

Weeds may reduce biodiversity by displacing populations of more diverse native species. Lantana (*Lantana camara*) is a serious weed in Australia and Africa that was originally a native of tropical Central and South America. The plant is relatively inconspicuous in the wild in Mexico except for the attractive flowers. The plant was introduced into cultivation as an ornamental in Europe and from there garden cultivars were distributed around the world, becoming naturalized as a very vigorous weed. In subtropical Australia and in similar regions of South Africa, Lantana has become a dominant species, colonizing any cleared areas and forming a dense understorey in Eucalypt forests and in gaps in rainforests. This example illustrates the problems created when humans move plants from there native environments and put them in new environments without the constraints to population growth (e.g. predators, pests and diseases) that applied in the environment in which the species evolved. Several insect pests of Lantana have been introduced into Australia in attempts to control this weed that have met with relatively little success.

Lantana is a serious weed in subtropical forests in Australia. This plant has been a major weed for more than 100 years. Lantana, a native of Mexico, was taken to Europe and developed as a garden plant before spreading to warmer climates worldwide, where it has become a major weed. Lantana, as shown in Figure 7.2, is a major weed of subtropical cleared areas (photographed in Australia) and (inset) has a wild relative growing in

Figure 7.2 *Comparison of a weed in a new environment and in its native habitat*

Mexico. Plants that are not dominant in the landscape of their native habitat often become major weeds in a new environment, away from factors controlling the population in the location in which they evolved.

Why conserve biodiversity?

We need to understand plant diversity and how to measure it, but we also need improved understanding of why plant biodiversity is important (Henry, 2005b). Plants are an essential component of ecosystems that provide a wide range of essential services on earth. These ecosystem services are outlined in Box 7.4.

Plant diversity is a direct contributor to ecosystem productivity. The greater the diversity of a biological system the greater the total biomass and positive environmental contribution the ecosystem can deliver. Different species co-existing together in a diverse plant community occupy different parts of the environment in space and time, allowing the maximum biomass to be supported and increasing the stability of the ecosystem. Diverse communities may be more robust and able to adapt to environmental change.

Box 7.4 *Ecosystem services*

Ecosystems provide a wide range of services that are of value to humans. Plants play a central role in many of these services.

Examples of the role of plants in ecosystem services include:

- Clean air: plants fix carbon dioxide and produce oxygen.
- Food: plants provide food directly.
- Waste decomposition and nutrient cycling: plants help create an environment for microorganisms that degrade organic matter.
- Climate: plants directly impact climate by transpiration, contributing to the water cycle.
- Clean water: aquatic plants provide a habitat for surface organisms that remove many solutes from water and can purify water for drinking.

Diversity of plants is also an essential resource for plant improvement in the development of useful plants for agriculture and forestry. Economic plants need to be continually genetically selected and improved to ensure they remain productive in domestication, and can be adapted to changing production environments or end-use requirements.

Biodiversity is very important to humans from an aesthetic perspective. Life is enriched by experiences or a wide range of diverse organisms. In contrast, a life in which it was only possible to encounter a very small number of other species would be less stimulating and desirable to many humans.

Option available for conservation of plant diversity

Collections aiming to conserve plant biodiversity *ex situ* (in a location other than the wild) include botanic gardens and seed banks. Private gardens are also an important site of *ex situ* plant conservation. Humans have long been obsessed with plants as ornamentals. Private plant collections in gardens remain an important contributor to the conservation of plant biodiversity. Plant collectors have not always made a positive contribution to the conservation of rare species. Wild populations of rare plants, especially plants considered attractive by humans, such as orchids, have suffered from over collection of wild populations. Indeed, any very rare plant is threatened by the attractiveness of rareness itself with the determination of many plant collectors to have rare species in their collections.

Figure 7.3 *Biodiversity in cultivation – traditional village gardens near Tsukuba, Japan*

The locations of rare plant populations are often not revealed in public databases to avoid the threats to the population that might result from ready access to details of population locations if they can be located by a search of the internet.

Domestication has created many new plant types that do not exist in the wild. The diversity of cultivated plants has become a very important biological resource. Gardens are an important contributor to biodiversity in many areas. A vegetable garden in a village in Japan is illustrated in Figure 7.3. The conservation of diversity of domesticated crops by farmers managing the genetic stocks that they cultivate is a major contributor to conservation of crop diversity.

Options for the conservation of both wild and cultivated plants *ex situ* include the use of living field collections (e.g. botanic gardens), seed banks, and the storage of pollen and DNA.

These are key technologies as the potential for *in situ* conservation is exhausted in specific regions (especially urban areas) due to human activities. The effectiveness of these approaches is critically dependent on effective collection strategies to ensure the diversity of remnant wild populations as represented in the *ex situ* populations.

Botanic gardens

Botanic gardens are key locations for conservation of plant diversity at local and international level. Living collections in botanic gardens may be linked to related collections in herbaria and to seed and DNA banks. Botanic gardens are widely distributed globally. Many have multiple roles in providing public open space (as public parks) and struggle to contribute to their other missions, including biodiversity conservation and community education. Many focus on collecting the plants of the region or location in which they are based. Others have attempted more extensive collections of plant diversity. Frequently these two objectives are combined. Increased levels of collaboration between public and private gardens using improving internet facilities may continue to enhance their potential to support plant biodiversity conservation on a global scale.

Box 7.5 *Lismore Rainforest Botanic Gardens*

This is an example of plant biodiversity conservation in a botanic garden being established on a degraded site. I have served as a member of the Lismore Rainforest Botanic Gardens Committee. This garden is being established by the city of Lismore to represent the subtropical rainforests of eastern Australia that were the original flora of the local area. The original rainforest, known locally as the 'Big Scrub', is now represented by very small patches of remanent vegetation. The original forest was probably the largest area of relatively continuous lowland subtropical rainforest. The forest was cleared for timber (especially the local red cedar, *Toona ciliata*, a member of the mahogany group (Meliaceae)) and then agriculture (mostly dairying) more than 100 years ago. This is a long-term project involving extensive community participation. The Australian Plants Society, essentially an organization of private gardeners, formed to promote the cultivation of Australian plants, has supported the project. The gardens are being established on a degraded site that has been used for the disposal of city waste as landfill. This provides an opportunity to demonstrate the potential for ecological restoration of degraded areas. Community education and engagement in the process are key objectives. The outcome will represent the plant biodiversity of the subtropical rainforests, delivering a direct conservation outcome and providing for ongoing educational opportunities.

Seed banks

Seed banks mainly focus on plants of economic species, but have the potential to be expanded to also conserve wild species more generally. Many collections extend to wild relatives of economic species (see Table 2.5), but not to other species without current economic value. Seed banks are often backed up by sharing of seed batches with other seed banks, often in another country. The Global Seed Vault in Svalvard, Norway is a permanent back-up system for seed banks. The location ensures that the seeds are maintained at low temperatures with the requirements of energy (electricity) at most other locations. This facility is supported by the Global Crop Diversity Trust. An international treaty on Plant Genetic Resources for Food and Agriculture has been established in harmony with the Convention of Biodiversity. This treaty provides for multilateral agreements that guarantee benefit sharing and the international exchange of essential plant genetic resources for the most important food crops. This process of extensive sharing has been driven by the importance of these collections, especially for species for which the conservation of wild or cultivated populations cannot be assured. This makes it difficult to assess the numbers of distinct seed lots (genetically distinct accessions or collections) in international collections. Large numbers are replicates of those held in other collections so that the total number of accessions in all seed banks internationally may seriously overestimate the extent to which the wild or cultivated diversity of the species has been sampled.

Seed banks generally rely on using low temperature and low moisture content to preserve seeds for long periods. They are usually not an option for seeds that do not survive desiccation (drying). This is a characteristic of many plants from wet environments such as rainforests. Living collections of the plant in a botanic garden or a dedicated field plot are the important options for these species.

Pollen storage

Pollen storage techniques have mainly been perfected for species that are the subject of active plant-breeding programmes. More widespread conservation of pollen would be an important aid to plant biodiversity conservation.

DNA banks

Management of *in situ* populations is refined by ongoing genetic research that is now mainly at the molecular (DNA) level. Understanding the impact of habitat fragmentation on the long-term evolutionary potential of species is essential to developing strategies for managing the islands of biodiversity that are often created by human activities such as urban, industrial or agricultural developments.

Advances in genomics science are making much more data available to researchers. These advances have been driven largely by medically focused research, but promise to provide much more powerful tools for monitoring and analysis of genetic diversity. The availability of DNA collections representing the diversity of plant populations will be a key factor limiting rapid applications of this technology in plant conservation. Plant DNA banks are only at the very early stage of development in most locations and require significant further investment to be able to fulfil their potential role in biodiversity conservation. For example, the Australian Plant DNA bank aims to hold DNA from every Australian plant species and to represent as far as possible the diversity within each species. It is a very long-term goal to collect DNA from all 20,000 or more Australian species. For very rare species it is possible to collect DNA from every known individual of the species, while with species with large populations sampling to represent the diversity becomes more challenging. A structured collection including all known distinct populations of the species, or covering the known geographic range of distribution of the species, is often attempted. Plant DNA banks are a relatively new concept complementing the seed and germplasm banks that have been established for much longer. An international network of plant DNA banks with these objectives is starting to emerge. with recent initiatives in Brazil and South Africa (see Box 2.5).

Novel techniques such as tissue culture of cryopreservation may be required for the *ex situ* conservation of some plant materials.

Tissue culture

The growth of plants or plant parts in tissue culture is a technique that has been developed for the commercial propagation of plants that are difficult to reproduce from seed or by conventional vegetative methods (e.g. cuttings). The conservation of specific rare genotypes or species may be greatly assisted by mass propagation using tissue culture methods. These techniques have been widely applied to high-value plants (e.g. ornamentals such as orchids), but have had only limited application to conserving biodiversity in other species, largely because of the cost. A major cost is the effort required to develop an appropriate culture method tailored to the specific nutritional requirements of each species or even genotype. Low-cost generic protocols that could be applied to many species would be widely adopted and remain a desirable research objective.

Cryopreservation

The long-term storage of plants or plant parts by freezing may allow conservation of rare genotypes until new technologies for propagation are developed, or more suitable environments are found for their establishment in the wild. This technique has had limited application to date, mainly for highly endangered material. The key limitation is that techniques for freezing and recovery of viable material have not been developed for many species.

In situ conservation

The availability of improved methods for the conservation of plants in environments created by humans (*ex situ*) should not be used as an excuse for reduced efforts to conserve plants in their native environments (*in situ*).

The creation of nature reserves and their protection has been a primary approach. These programmes need to be intensified in many areas. National parks face increasing pressures from growing human use and climate change.

The conservation of plants on private lands and in disturbed areas is also a very important strategy. Developing cooperation between landowners in a local area is an important element of achieving conservation outcomes as plants are not constrained within land boundaries imposed by humans. The Australian Landcare movement has proven to be a very useful forum for discussion of conservation issues between neighbours. This scheme encourages landholders in each area to join together, often on a catchment basis, to form Landcare groups to mange their local environment. These groups have become an important social network in many rural areas, possibly explaining their high participation rates and success. Landcare groups work on both public and private land, and support the protection of endangered species and coordinates forest regeneration efforts, especially the removal of weeds as a community activity.

How do we measure progress in conservation of biodiversity?

Unfortunately we currently have a lack of objective methods to measure our success in conservation of biodiversity. This has been highlighted in a recent study of progress in conserving areas of the state of Queensland (McDonald-Madden et al, 2008). Land clearing and reservation of land to conserve biodiversity are opposing activities and quantitative analysis of these competing factors is required to identify if progress is being made in conservation efforts. We currently do not have enough data and careful analysis to allow reliable assessment of biodiversity status and trends in most parts of the world.

The emergence of the threat of climate change is a major factor with potential to adversely impact upon biodiversity. This is the topic of the next chapter.

8

Impact of Climate Change on Biodiversity

Understanding the ecological importance of biodiversity for ecosystems functioning and ecological services to mankind requires us to relate the diversity of ecosystem properties to the diversity of species performances in space, in time, in biotic interaction and under changing environmental conditions.

Beierkuhnlein and Jentsch, 2005

Climate change and the rate of loss of plant species

The implications of climate change for biodiversity conservation (Lovejoy, 2006) are central to management strategies used in conservation of plant diversity. The importance of plant biodiversity to ecosystem functioning and as a resource for human food, feed, fibre and energy has been well illustrated in preceding chapters. The general acceptance of climate change as a major issue facing human societies was outlined in Chapter 3. Climate change is likely to accelerate the rate of species extinction in the immediate future (next few decades). Rapid climate change has resulted in loss of species diversity in the past. Many species are not able to adapt rapidly enough to environmental changes and will be lost. Estimates of the likely rate of species loss in the immediate future due to climate change vary widely. However, it is likely that the current high rates of species loss due to human removal of habitat are likely to increase with climate change. Very high levels of extinction of plants and animals are predicted by many models (Thomas et al, 2004). Food security long term may be threatened by loss of populations of wild relatives of crop species due to climate change. This will reduce options to continue to adapt crops to new climatic conditions.

Climate change and the rate of loss of plant diversity within species

The loss of plant species may be relatively easy to monitor. However, changes in climate will also result in some genotypes within the population being favoured. This loss of genetic diversity within species may not be easily detected even when serious loss of diversity has occurred.

Recent studies have examined the genetics of wild barley populations in relation to differences in climate (Box 8.1; Cronin et al, 2007). This area requires much more research. Advances to technologies for plant DNA analysis promise to make this type of research more cost effective in the future. This will hopefully allow more detailed analysis of diversity within species, provide the tools required to monitor the impact of climate on diversity and the opportunity for more active management and conservation of biodiversity.

Box 8.1 *Impact of climate on wild barley populations*

Wild barley remains extensive today and represents the populations from which our first crop plant was probably domesticated. Analysis of the DNA from wild populations has been used to examine the links between gene diversity and climate in different populations. This system was introduced in Box 3.2 in relation to identification of genetic options for adapting food crops to climate change. Wild barley extends over sites that vary in a wide range of climatic variables – from rainfall and humidity to temperature extremes and averages. The soil type may also be important in determining the amount of water retained in the soil and the resulting level of water stress faced by the barley plants. The diversity in a gene associated with expression of a protein in the seeds of the plant that apparently provides defence against fungal pathogens (probably in the soil) varied with the environment in which the barley populations were found. Populations from dry environments had greater diversity in this gene than populations from wetter environments. These observations can be explained by the presence of more diversity in the populations of fungi in the soils of dry environments. The wetter environments may favour fungal growth but the fungi present are less diverse. More recent research has confirmed this finding for other genes and identified different patterns of expression for genes with different functions in the plant.

Conserving plant species *in situ*

The preferred first option for the conservation of plant biodiversity is to protect them *in situ* (where they are in nature). This will continue to be the first option explored. As this becomes more difficult, a move to more efforts in *ex situ* (in an artificial location – e.g. botanic garden, seed bank, greenhouse) conservation will become essential. A future in which we can only experience plants in a garden or an artificial environment is one in which we concede defeat in the fight to protect wilderness areas (with all their associated human emotional and aesthetic values), but realistically may become the only option in many cases. Because it is the best way to conserve entire ecosystems with all their diversity, *in situ* conservation must remain a high priority. As already emphasized, we should not allow the development of *ex situ* conservation technologies (e.g. cryopreservation or tissue culture), no matter how effective they are, to become an excuse for reduced efforts at *in situ* conservation.

A mixture of protected areas and strategies to conserve diversity outside these areas is required to conserve biodiversity for different types of species. The designation of protected areas is a key strategy for conserving plant biodiversity *in situ*. Many rare plant species are only found in very small areas with unique habitats. Protection of these areas is probably the only realistic option for conservation of these species, but significant climate change could lead to extinction of these species. Outside protected area policies are required to encourage the management of landscapes to allow a mosaic of nature conservation and agriculture. Areas of land reserved for nature conservation within agricultural landscapes can improve sustainability of agricultural production on the parts used for agriculture. They are also essential refuges for these animals, and corridors for the movement of animals and even the dispersal of plants. Habitat is required to support the insects' need for crop pollination. Trees along field margins may assist the sustainability of agricultural production by lowering the water table in the adjacent field. This can be critical when subsoils have factors such as salt that threaten crop performance. These approaches are key strategies for achieving environments better able to adapt to climate change.

Box 8.2 *Climate change and* Banksia conferta

Banksia conferta, the rare *Banksia* on the peak of Mount Tibrogargon in Queensland (Figure 8.1), may be threatened by climate change that results in more frequent fires. In recent years fires have swept across the mountain top in successive seasons. Many species including *Banksia confera* are killed by fire, but regenerate readily from seed following a fire. Indeed, fire is often necessary for long-term continuation of the species. Frequent fires, however, may pose a significant threat. A fire that kills seedlings in successive years before the plants have reached an age at which they flower and set seed will deplete the seed reserves in the soil. Widespread species can cope with this if the entire population is not burnt. However, on the small mountain peak a fire can kill the entire population. Survival of the species depends upon the seeds in the soil being able to grow to maturity and set more seed before the next fire. The high frequency of fires may be a result of a greater density of human settlement surrounding the mountain, but may also be a consequence of climate change.

Figure 8.1 *The rare* Banksia conferta. *This population is growing on the peak of Mount Tibrogargon, Australia*

Relocation of species

Some species may be able to adapt by moving to an environment to which they are adapted. For example, species distributions may move to higher altitude as the temperature increases. In this way they may be able to exist in an environment that has similar temperatures to the one they were adapted to before the temperature rose. This has limitations as species have nowhere to go when they reach the top of the mountains. Alternatively their distributions may move to higher latitudes to stay within the temperature range in which they evolved. Unlike animals, individual plants cannot move. However, movement of populations can take place over generations. Species with a short generation time (rapid development to reproductive maturity) and long-distance seed dispersal mechanisms are likely to relocate to favourable environments and cope better with rapid climate change than long-lived (slow to reach reproductive maturity) species with a limited seed dispersal range. Unfortunately these responses are now constrained in many areas because of human development or land use. Many species are located within limited reserves (sometimes created to protect the species) that have become islands in an ocean of urban or agricultural development. Wildlife corridors have been created largely to allow animal movements, but could also be important for plant movement in these situations. Human intervention in the form of deliberate translocation will become increasingly important to ensure the survival of many such plant species. Creation of new reserves to cope with this problem will also become necessary. More research on methods for effective plant translocation is required so that we have available technologies for the rescue of species. Many attempts at translocation of rare plants, especially large trees, have been unsuccessful. Accelerated movement with human assistance may create other environmental problems and probably has a low chance of success (Hoegh-Guldberg et al, 2008).

Conserving plant biodiversity ex *situ*

The *in situ* conservation of many species will prove impossible despite our best efforts and we will need to increase the scale of *ex situ* conservation efforts. Technologies for *ex situ* conservation will need to continue to improve to provide for the efficient achievement of *ex situ* conservation. Rapid climate change may demand an acceleration of the development and application of *ex situ* conservation measures. More extensive botanic gardens, seed banks and DNA banks may be required, and they may need to be allocated significantly more resources to cope with increasing demand to conserve genetic diversity of many plant populations.

Impact of the use of plants for energy (fuel) on plant biodiversity

The production of biofuels from plants may be a response to climate change, and a key approach to reducing the levels of greenhouse gas emission and reducing the risk of climate change to biodiversity. However, biofuel production may itself become a threat to plant biodiversity. The use of plants for biofuel production will require land and this may be at the expense of areas that could be used for biodiversity conservation. However, if this use makes a significant contribution to reduction in the rate of greenhouse gas emissions, then it may have a net positive effect on biodiversity by limiting the rate of climate change and giving more species a chance to survive. This should be a key criterion for deciding if use of plants for biofuels is acceptable. The life-cycle assessment of the biofuel production system should provide evidence of this net benefit. This issue will be taken up in Chapter 9.

9

Competition between Agriculture and Biodiversity Conservation

Moreover, had more environmentally fragile land been brought into agricultural production, the impact on soil erosion, loss of forests, grasslands, and biodiversity, and extinction of wildlife species would have been enormous.

Norman Borlaug (2004)

The competition between human needs for land for agriculture to meet our requirements for food and our desire to live in a biologically diverse environment has been an issue that has often been avoided. We all need to eat and we (mostly) appreciate the aesthetic value if not always the survival value of biodiversity. The conflict between these two key human drivers is not resolved unless we understand the competition between these worthy objectives at the level of competition for space on a planet that is increasingly crowded by human activities. The emergence of energy as another competitor in this space might help to focus us more on resolving the central issues of balancing the immediate day-to-day needs of humans and their aspirations for long-term survival on earth. This chapter will examine the competing needs of growing demand for agricultural products and the need for biodiversity conservation. We can and do make active decisions to limit our agriculture and to conserve areas for biodiversity. Japan provides an important example of a large population in a developed country that has retained a very high proportion of the country (more than 70 per cent) as forest. This has been the result of long-term policies. The consequences for the environment and biodiversity have been very positive, but Japan imports much of its food.

Biodiversity implications of food or fuel production

Sufficient agricultural land may be found to produce food and fuel, and balancing these two requirements has been discussed in Chapter 6, but at what cost to biodiversity conservation and the other essential functions of plants in the environment? Plants are essential to the global carbon cycle and generation of oxygen in the atmosphere. However, these roles can also be contributed to by crop plants. The essential role of non-crop (non-economic) plants is in biodiversity conservation generally. The replacement of natural vegetation with crops producing biofuels may result in a large emission of greenhouse gases that is very much greater (17 to 420 times – Fargione et al, 2008) than the annual reduction in greenhouse gas emissions that would be contributed by the biofuel crop. Replacement of tropical forests with biofuel crops is likely to be undesirable from both a climate and biodiversity perspective (Danielsen et al, 2008). In contrast, the use of waste or growth of perennial biofuel crops on degraded or abandoned agricultural lands carries little carbon debt and may provide very significant benefits in reduced greenhouse gas emissions.

A good example of the issues that need to be addressed is provided by palm oil production. Palm oil is a source of food oil and also a potential biodiesel crop. The production of palm oil has become controversial, with opposition to the expansion of production in countries such as Malaysia, Indonesia and New Guinea on the basis that areas of high biodiversity value are being displaced for the planting of oil palm plantations.

A contrasting example may be provided by the opportunity to grow perennial biofuel crops such as Miscanthus and Willow in environments such as the UK. These crops are fast growing and because of their perennial nature the energy of greenhouse gas costs of production are low. Displacement of annual crops with these species would represent a major change in the rural landscape (Karp et al, 2009). The implications for biodiversity are not well understood but are probably positive.

Objective analysis of the economic, environmental and social implications of food or energy production is required to determine the merits of different food or energy crops and their likely impact. This is best achieved by LCA examining the total economic, energy, environmental or social impact. The net energy balance and net greenhouse gas emission reduction of biofuel production systems can be used to compare alternative options (Hill et al, 2006).

Policies to promote desirable outcomes

Public policies include carbon trading, tax and mandating biofuel or renewable energy targets, and can be a key driver of change. Developing policies that are sound from an economic, environmental and social perspective is challenging. The wrong policies could easily deliver undesirable outcomes. Mandating a level of biofuel inclusion in transport fuels will probably result in production of the target, but we may not be pleased if this is achieved at an undesirable social or environmental cost. In developing policies, we can evaluate technologies as they exist today. However, there are great efforts under way to invent a new future. The rate of technology advance and more critically the directions that might take us in terms of the best options to pursue is almost impossible to predict. This suggests that our approach should be subject to constant review and updated rapidly to respond to developments.

Policies need to ensure sustainable biofuel conversion technologies. For example, the amount of water needs to be kept to a minimum by using water recycling approaches. However, biofuel production may already be much more water-efficient than gasoline production from oil. The greenhouse gas balance is favourable even for first-generation biofuels when compared to fossil fuels and will be dramatically improved by the development of second-generation biofuels (Hill et al, 2006).

Policies need to be set so as to encourage environmentally positive outcomes in any changes in agricultural production for food or energy. Factors that need to be considered as policy targets include:

- greenhouse gas balance;
- energy costs;
- nutrient run-off;
- hydrological impact;
- biodiversity impact (both within the crop and adjacent);
- productivity (high yields reduce the land footprint of agriculture).

The role of food and energy prices

High food prices may encourage the clearing of more land to produce food at the expense of biodiversity. Food prices have been historically low until recently. Oil prices are likely to rise as supplies are exhausted. As technology improves the cost of biofuels has been reduced. The cost of biofuels relative to fossil fuel-derived fuels should continue to decline. Transport fuel costs are likely to remain a significant component of food prices; indeed, some recent rises in food prices are probably largely due to increases in oil prices. Competition for land and water could put upward pressure on both food and biofuel costs. This has become a focus of public concern. However, the alternative to continuing to use fossil fuels will probably be much more expensive both for energy and for food (also with embedded energy costs). Many will argue that this makes the case for pursuing other alternative energy options, but these may not be technically feasible on the timescale needed to avoid unacceptable global climate change. The total of energy and food costs together needs to be minimized to be politically acceptable. Food and energy costs may need to rise relative to the prevailing levels of recent decades, but this may be a necessary part of our adapting to the threats we now face. Higher food and energy costs could have desirable impacts in affluent developed countries by providing an incentive to reduce consumption. The problem is really one in developing countries, where it will be important to protect the poor from shortages of affordable food that could result. Addressing this issue globally is an important step in adapting to the future.

Crops as weeds (especially new biofuel crops) impacting adversely on biodiversity

Weeds usually impact negatively on biodiversity. A weed by definition is a plant that is not wanted in the place that it is found. A more important issue is that it is has not evolved as part of the ecosystem in which it is now established. The usual consequence of a weed is the displacement of other species native to the area, sometimes to the point of threatening endemic species with extinction, but often at least contributing to a reduction in biodiversity by displacing a diverse local biota with one dominated by large populations of the weed species. One of the risks that needs to be managed is the potential for new energy crops to become major weeds. The desirable characteristics of an energy crop, that allow rapid growth on marginal lands, are those associated with species that have potential to become weeds. Harvesting of weeds for biofuel production has been suggested as an economic way to control some woody weeds. The Camphor Laurel tree (*Cinnamomum camphor*) from China has become a major weed in the high rainfall areas on the coast in central eastern Australia. These areas were mainly subtropical rainforest with a very high biodiversity that were cleared for crops and then pastures, and have been colonized by the Camphor Laurel trees to the exclusion of almost all other plant species in some areas. Economic use of this species may support its harvesting, especially in areas where it has become the main tree species, and this may allow regrowth of a wide range of species contributing to enhanced biodiversity.

Extent of competition (for land water and other resources)

The competition for land depends upon population growth and the resulting demands for food and fuel. The estimates of population growth and food supply and demand discussed in Chapter 6 have been used to calculate some

Table 9.1 *Land requirements to satisfy food and fuel requirements by 2085 – relative to 2005 values*

Population	Diversion of land to biofuels	Extra land required
Low	None (0%)	–24%
(10.5 billion)	Intermediate (30%)	9%
	High (50%)	53%
High	None (0%)	7%
(14.7 billion)	Intermediate (30%)	52%
	High (50%)	114%

scenarios for the land use requirements for food and fuel production to 2085 (Table 9.1).

The high population option seems highly undesirable, especially if we require a high portion of the land for biofuel production. In the lower population scenario, demand for land only becomes a significant issue if we have a high proportion devoted to biofuels. Expansion of agricultural activity by 50 per cent is an outcome of high biofuel production at low population growth and intermediate production at a higher population level. This is probably the type of outcome we should consider in terms of likely impact on biodiversity. However, we probably cannot afford to use 50 per cent more land for agriculture globally. Some re-prioritization within the land already used or assigned for agriculture will be necessary.

The expansion of sugarcane production in Brazil, as discussed in Chapter 6, is a good example of the issue. Many people in the world are aware of the rainforests of the Amazon and expect these forests to be a major area of conflict between agricultural (food and feed) production and conservation. However, in this example, other parts of Brazil may present examples of greater competition for land use. The rainforest is difficult to clear and in some ways less threatened than other communities. The Atlantic forests in Brazil are much more critically endangered because of population pressures along the coast. The Cerrado is a large area of savannah (more than 2 million km^2) with more than 10,000 plant species, more than 40 per cent of which are endemic. This is the area that is more likely to be used for expanded sugarcane production to satisfy energy demand. The desired approach in this area is to utilize land that has already been cleared for agriculture – mainly degraded pastures.

The latest analysis (2006 state of the environment report) of land use in the state of New South Wales (Australia) indicated that 70 per cent of land is used for grazing, less than 8 per cent is in protected areas, and a similar amount, around 8 per cent, is used to grow crops. Forestry occupies less than 4 per cent and urban and mining areas make up only 0.3 per cent of

Table 9.2 *Estimates of yields, conversion efficiencies and areas of land required to replace oil with biofuel in a country consuming around 100GL/year fuel consumption*

Yield (tonnes/ha/year)	Proportion of biomass converted to biofuel (%)	Land area (mha)
5	25	80
	50	40
	75	26
10	25	40
	50	20
	75	14
20	25	20
	50	10
	75	7
50	25	8
	50	4
	75	3

the state. This indicates the main area for reallocation to energy crops would be from the very large area devoted to grazing. Similar conclusions may be reached in many other regions.

Critical questions become just how much pasture can we convert to crop production? What types of vegetation (and associated biodiversity) are currently found on the land that would need to be used for crops? This needs to be assessed on a regional basis. The impact and the issues in Europe may be very different to that in Africa or South America. Some changes in land use could be positive for biodiversity if crops that provide a suitable habitat for wildlife are planted in degraded environments. Integration of food and energy production on farms could be based upon a greater diversity of crops and a much greater standing biomass, especially if trees are included in the plants farmed. The next chapter will explore our option for domesticating new plants for these applications. The choice of species for biofuel production and the land footprint required depends upon the biomass yield per hectare and the proportion of the biomass that can be converted to fuel. The land requirements for replacement of oil use with biofuels in a country around the size of Canada are outlined in Table 9.2.

This analysis indicates that combinations of large areas of low yielding crops (or crop residues) and smaller areas of high yielding crops (dedicated energy crops – e.g. trees or energy grasses) could be used. The cost of harvesting and transport will make the more intensive options with the smallest land footprint more attractive. However, these options will probably require the use of better land with more chance of competition with food production. While this needs to be avoided as much as possible, the small footprint and the potential environmental benefits of replacing fossil fuel use may

Table 9.3 *Water use for electricity and biofuel production from different crops*

Species	Water footprint (m³/GJ)*		
	Electricity	Bio-ethanol	Biodiesel
Sugar beet	27	59	
Maize	20	110	
Sugarcane	27	108	
Barley	39	159	
Rye	36	171	
Rice	31	191	
Wheat	54	211	
Potato	47	103	
Cassava	21	125	
Soybean	95		394
Sorghum	78	419	
Rapeseed	229		409
Jatropha	231		574

* Weighted average water footprint – volume of water required per unit of energy produced – data from Gerbens-Leenes et al, 2008

justify this choice. Giving up a small area of land to reduce the risks of climate change may be a net benefit to food production.

The competition for land is not the only basis for competition. Plant growth requires other inputs including water and a wide range of nutrients from the soil. Water is probably the most critical of these, especially in the context of potential reductions in rainfall in many regions due to global climate change.

Analysis of the water-use efficiency of food and energy crops can be used to compare cropping options. Recent analysis of biofuel crops (Gerbens-Leenes et al, 2008) has emphasized the wide differences in water requirements of different production systems (Table 9.3). The production of electricity is more efficient for these crops because it is based upon using the entire biomass. Biofuel production efficiencies in Table 9.3 are based upon first-generation production of ethanol from sugars and starch or diesel from seed oils. The poor water-use efficiency of these processes, especially the production of biodiesel, is emphasized by these estimates. Biodiesel production from Jatropha required almost ten times as much water per unit of energy produced as bio-ethanol production from sugar beet. This analysis demonstrates the importance of development of second-generation biofuel crops that allow all of the biomass to be converted to fuel energy.

The relative importance of land and water use needs to be accessed on a local or regional basis. For example, in a desert water use is critical, but land may be readily available and in a very high rainfall area water may not be limiting, but conservation of land for diverse rainforest communities might be very important. The growth of crops that are adapted to the local environment is probably the key desirable objective to ensure optimal use of land and water resources and to minimize competition with biodiversity. Analysis of the land available for biofuel production in the UK has identified 3.1 mha of suitable land (Haughton et al, 2009). This area was predicted to be environmentally acceptable. Planting of crops such as Miscanthus and willow was likely to increase biodiversity in many areas relative to that associated with current food crop production.

Direct impact of the type of vegetation cover (native vegetation or agriculture) on climate

The impact of agriculture on the albedo of the landscape may be an import issue if land use patterns are altered over large areas of land. The albedo is the diffuse reflectivity determining the percentage of incident solar radiation that is reflected from the land surface. A bright surface such as snow reflects back much of the light, while a dark surface such as a forest or a crop may not. Land use changes that influence albedo can contribute to warming. Recently the development of crops that have a higher albedo has been proposed as a strategy to combat climate change (Stephenson, 2009). For example, corn with a high level of surface waxes could be bred to reflect more light and lower temperatures. The management of land use to ensure the vegetation cover does not adversely impact in this way needs to be balanced against the impact of the changes on greenhouse gas accumulation.

The importance of new technology

Better tools for assessment of land use and biodiversity monitoring are needed to allow better management. We cannot manage what we cannot or do not measure.

Technology development is an essential requirement for coping with demands for food and energy while conserving biodiversity. The competition between agriculture for food and energy and biodiversity will be greatly reduced if second-generation biofuel crops are produced. More efficient agriculture (higher production per unit area) is a key requirement to minimize the pressure for more land for agriculture. This will require the continued advancement of all areas of agricultural and plant science.

Need for continued reservation of land for nature conservation

Conservation of areas of high biodiversity remains critical. Rainforests of eastern Australia are illustrated in Figure 9.1. The identification and protection of such areas will be an ongoing requirement, especially in the face of competition from agriculture for land use. The most difficult areas to ensure continuing protection are those remaining areas with unique biodiversity in areas of high agricultural value.

Reserving specific areas for nature conservation and others for agriculture is an important mechanism for resolving the balance between these two requirements, but some level of integrated agriculture and conservation in

Left panel: Subtropical rainforest (Nightcap Ranges, northern New South Wales). Right panel: Tropical rainforest (near Kuranda, North Queensland). Both of these rainforests are protected as world heritage areas. These two well-separated areas (more than 1000km apart) contain related but distinct species in some groups. For example, *Hicksbeachia pinnatifolia* (Figure 10.4) is found in the southern sub-tropical rainforests, while the related *Hicksbechia pilosa* is found in the northern tropical rainforests. Similarly, *Davidsonia jerseyana* is found in the subtropical regions and *Davidsonia pruriens* (Figure 10.5) in the tropical areas. In both cases the differences are slight but significant enough to regard the plants from these geographically well-separated forests as distinct species. Competition from agriculture reduced the area of these forests more than 100 years ago. The remaining areas require ongoing conservation to ensure this biodiversity is not lost.

Figure 9.1 *Rainforests*

the same landscape is required. Some of the advantages to agriculture of biodiversity in the landscape have already been discussed in the last chapter. More innovative approaches to management of agricultural areas may be required to promote biodiversity within the production environment (Perfecto et al, 2009).

Conclusions

Recent analysis (Metzger and Hüttermann, 2008) suggests that sufficient biomass to support human transport and other energy requirements could be produced sustainably by the growth of trees on land that has been degraded by human activity. This should be the priority to avoid competition with food production. Biodiversity conservation in these areas will be a key constraint to manage. Degraded areas may have significant biodiversity values if they support patches of remnant vegetation.

We currently use around 40 per cent of the land surface of the Earth for the production of crops and pastures (Stokstad, 2009). Increasing efficiency of agriculture (more food per unit of area) is an essential requirement if we are to avoid expansion into areas that would compete with nature conservation and contribute to a significant loss of biodiversity. The efficiency of agriculture has been improving rapidly in recent years, but we need to continue this trend and do so sustainably if we are to meet the challenges of future growth in food and energy demand.

10

Domestication of New Species

But whether or not the selection of wild edible plants by ancient hikers relied upon conscious or unconscious criteria, the resulting evolution of wild plants into crops was at first an unconscious process. It followed inevitably from our selecting among wild plant individuals, and from competition among plant individuals in gardens favouring individuals different from those favoured in the wild.

Jared Diamond (1997), *Guns, Germs and Steel*, p130

Domestication of new species has the potential to deliver outcomes that meet the challenges of efficient food feed and energy supply in response to growing human demands. However, the feasibility of domesticating new species has been challenged and needs to be evaluated. The key issue is whether science can define needs and opportunities that have been overlooked by the processes that have driven human domestication to date. Most plants have been domesticated for food use and domestication was introduced in Chapter 2. The emergence of new plant uses such as the provision of energy suggests the need to domesticate new species for these purposes. The modification of plants to improve their suitability as energy crops will require the selection of genotypes with novel traits that are not necessarily optimal for survival of the plant in the wild. For example, an improvement in accessibility to enzyme digestion of carbohydrates in plant cell walls would facilitate conversion to sugars and use in fuel production. This would potentially include steps to reduce the crystalline nature of the cellulose and to reduce lignin content. Fears that this would lead to plants that would be susceptible to pests and diseases have been expressed. While these unintended consequences need to be considered and managed carefully, comparison with the domestication of food species suggests that these concerns may not be impossible hurdles. Food plants have many characteristics that have been selected to suit humans and that would be deleterious to the survival of the plant in a wild population. Plants grown under domestication have human intervention to propagate them and to remove competitors, provide nutrients and water, and protect against pests and diseases.

History of domestication and implications for further domestication

The history of domestication of plant species to date should guide our approaches to the domestication of new species for new applications. The bottle gourd (*Lagenaria siceraria*) was domesticated very early as a storage vessel in tropical Africa and was taken by humans to America. This plant has a range of uses that suited the mobile hunter/gatherer.

Domestication of the major agricultural species has centred on a relatively small number of locations around the world where a critical number of species, with characteristics that suit them to domestication, are found in the same location. The best defined of these regions is the Fertile Crescent at the eastern end of the Mediterranean (Figure 10.1).

Barley was probably the first major food crop domesticated around 12,000 years ago, with wheat, peas, chickpeas and domesticated animals also originating in this area, all around 10,000 years ago. Evidence of early human cultivation of cereals is abundant (Figure 10.2). Botanical archaeologists like Dorian Fuller, University College London, are working to understand the relationships between plants and early human societies, before, during and after domestication (Fuller et al, 2007). Analysis of the genetics of early crops can reveal knowledge of great value in attempts to continue the domestication process or to accelerate it to meet new or growing human needs. The wild populations and landraces (traditionally cultivated cultivars) have many characteristics that are important resources for ongoing genetic improvement of modern cultivated crops. Domestication of genes from wild populations continues. Berhane Lakew, a barley breeder from Ethiopia, has recently defined genetic sources of drought tolerance in wild barley that have potential use in developing drought-tolerant crop cultivars. Abderrazek Jilal, a barley breeder from Morocco, has also recently characterized the genetics and human food qualities of wild populations and landraces as a basis for producing more food directly from barley crops. These important research projects have been conducted in collaboration with the International Center for Agricultural Research in the Dry Areas (ICARDA) in Allepo.

This is not a highly productive agricultural environment and many of the more productive environments in the world do not seem to have produced any domesticated plant or animal species. Diamond (2005a) offers many possible reasons for the success of domestication in the Fertile Crescent:

- Grasses with large seeds were abundant in this region (including the progenitors of wheat and barley) because the climate suited a species that devoted a lot of energy to the production of large seeds that could survive the hot dry summer and grow during the wetter and cooler months.

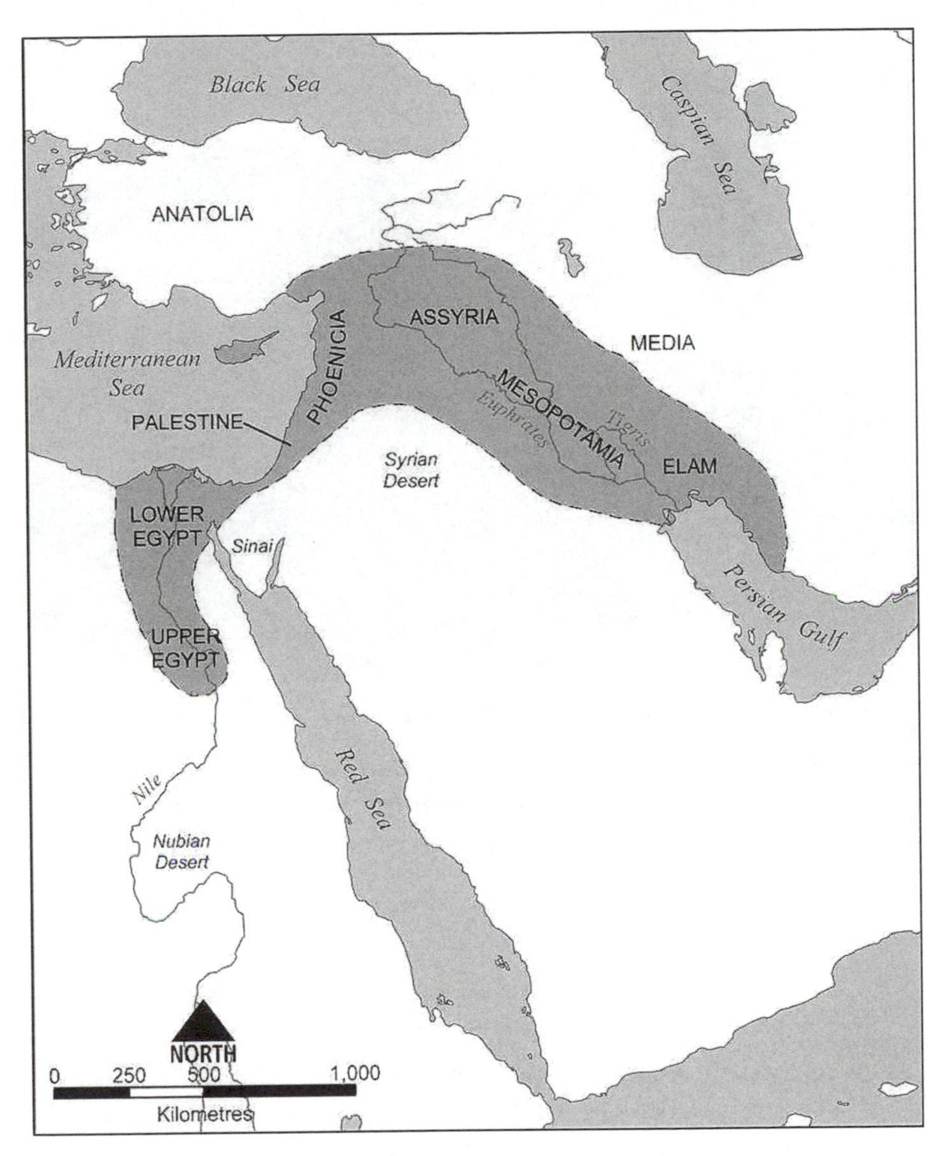

Figure 10.1 *Fertile Crescent – site of early domestication*

- Pulses (grain legumes), including pea, lentil and chickpea, were also domesticated in this region. The combination of cereal and pulse provides both the carbohydrate and the protein for a balanced human diet.
- Large mammals were also abundant in this region. The sheep, goat, pig and cow were all domesticated in this region. They also lacked the predatoriness of mammals in similar open grasslands of Africa, resulting in behavioural characteristics that made them better suited to human interaction.

Photographs provided by Dr James Helm (Field Crop Development Centre, Lacombe, Canada)

Figure 10.2 *Ancient grain. Samples of barley (inset) collected in the 1950s by Dr Robert Metzger in Turkey (fortress dating from around 2800 years ago)*

Box 10.1 *Did plants and animals domesticate humans?*

The traditional human-centric way of considering domestication is to think of humans domesticating plants. However, in an evolutionary or biological sense it is also worth considering these developments from the perspectives of the plants and animals. Did barley plants evolve to become so attractive to humans – the primate, *Homo sapiens* – that the humans assisted the reproduction of the plant by collecting the seed and planting it in the soil and even killing competing plants? In turn some of the plant population was eaten by the humans. The evolution of the plant into a domesticated form involved selection (by humans) for genes controlling domestication traits that provided these barley plants with a key adaptive advantage (under the conditions of human cultivation) over individuals in the population without these traits.

- This region is at the centre of the largest land mass on earth with great potential for plants to be transferred east and west, allowing the spread of agriculture. Plants generally do not perform well if they are moved north or south because of sensitivity of growth and flowering to a day's length.

Domestication may have been preceded by a period of pre-domestication in which humans combined wild harvest and some limited cultivation. Early harvest from cultivated grain may not have immediately met all food needs, resulting in supplementation of this with grain harvested from wild stands. This would be especially true in years of crop failure. The archaeological records suggest that wheat and barley underwent a prolonged period of pre-domestication in which the grain size was increased by human selection, before shattering was selected against, to give truly domesticated cereals (Fuller, 2007). This contrast with rice in which loss of shattering is observed at a similar time to or before an increase in grain size. The time to domesticate has been estimated by experiments in which stands of grain have been harvested to simulate wild harvest and the impact on domestication traits analysed. These experiments suggest short domestication periods of the order of 100 years. The domestication of cereals may have taken more like 1000 years. One explanation for this is the combination of wild harvest with cultivation. If the seed from cultivated material and wild harvesting are combined the impact of genetic selection under cultivated conditions is diluted, slowing down the genetic change in the population and slowing domestication.

Box 10.2 *Parallel domestication of plants and animals*

The domestication of plants and animals may have proceeded together as humans became less mobile and adopted a more sedentary lifestyle. Despite their apparent great diversity, all domesticated dogs were apparently domesticated from the grey wolf in what was probably a single event in the Fertile Crescent somewhere in the period 15,000 to 40,000 years ago. Dogs may have coexisted with humans, travelling the same migratory pathways following migratory food animals. This would have provided an opportunity for long contact between dogs and humans to allow the process of behavioural change necessary to transform a wild wolf into a domesticated dog. Cats in contrast may have essentially domesticated themselves. As humans built grain stores that attracted rodents, cats could move in to feed on the new food source.

The selection for seed size and shattering resistance in the cereals was probably not a conscious process. However, the selection for size in fruit probably was a very conscious process, explaining the very rapid increases in fruit size that evidence from domestication suggests.

Some examples of domestication are listed in Table 10.1.

Table 10.1 *Domestication of crop plants*

Crop Cultivated species		Progenitor species (or modern relative of progenitor)	Domestication site
Barley	*Hordeum vulgare*	*Hordeum spontaneum*	Fertile Crescent
Maize	*Zea mays*	*Zea mays ssp. parviglumis*	Meso-America
Rice	*Oryza sativa*	*Oryza rufipogon*	Asia
	Oryza glaberrima	*Oryza barthii*	Africa
Wheat	*Triticum aestivum*	*Triticum monococum*	Fertile Crescent
		Triticum. speltoides (relative)	
		Triticum tauschii	
Sorghum	*Sorghum bicolor*		Africa
Sugarcane	*Saccharum sp*	*Saccharum officinarum*	New Guinea, South East Asia
		Saccharum spontaneum	
Banana	*Musa*	*Musa acuminata*	SE Asia
		Musa balbisiana	
Cassava	*Manihot esculenta*	*M. esculenta ssp flabellifolia*	Amazon
Potato	*Solanum tuberosum*	*Solanum* species	Peru
Lima bean	*Phaseolus*	*Phaseolus lunatus*	Andes
			Mexico
Olive	*Olea europaea*		Near East
Sunflower	*Helianthus annuus*		North America
Squash	*Cucurbita pepo*		North America
			Mexico
Grape	*Vitis vinifera*		Eurasia
Peanut	*Arachis hypogaea*		Amazon
Cotton	*Gossypium*	*G. herbaceum*	Meso-America
		G. arboretum	South America
		G. hirsutum	
		G. barbadense	
Common bean	*Phaseolus vulgarus*		Meso-America
			South America
Yam	*Dioscorea rotundata*		West Africa
Flax	*Linum usitatissimum*	*Linum angustifolium*	Near East

Box 10.3 *The domestication of rice*

Rice domestication involved the separate domestication of two species: *Oryza rufipogon* to produce domesticated *Orzya sativa* in Asia; and *Oryza barthii* to produce *Oryza glaberrima* in Africa. The history of domestication of rice is complex (Sweeny and McCouch, 2007).

Asian rices have long been separated into Japonica and Indica types by plant breeders and taxonomists, with Javonica being defined as a third group. Molecular phylogenetic analysis provides further insights into these major groups of rices. The following five groups can be defined:

- Indica
- Aus
- Tropical Japonica (Jarvonica)
- Temperate Japonica (Japonica)
- Basmati

The Indica and Aus groups are distinct from the other three which also form a related group. The Indica group is more diverse than the Japonica group. It is notable that the Basmati rices are related to the Japonica rices, not the Indica rices, despite the Indica and Basmati rices sharing long grain length. Exploring the origins of rice by analysis of wild populations in Asia is complicated by the likely extent of gene flow from cultivated rice into wild populations. Gene transfer between divergent rice germplasm is also suggested by genetic evidence (Kovach et al, 2007). The case for single or multiple domestications of Asia rice has been debated in the scientific literature. Recently, Duncan Vaughan and colleagues from the National Institute of Agrobiological Sciences in Japan have suggested that a single domestication explains available evidence (Vaughan et al, 2008).

Research on the pericarp colour trait (red in wild rice, white in cultivated rices) indicates that this probably arose in Japonica rice and was transferred by out-crossing into Indica populations. Similarly, the long grain trait may have arisen in Japonica types and transferred to Indica rices.

Recent developments that illustrate the potential for domestication of new rice types are the combination of the Asian (*Oryza sativa*) and African rices (*Oryza glaberrima*) to produce new options. Sourcing of genes from other non-domesticated *Oryza* species (see Chapter 2) is also in progress.

What happens to genetic diversity during domestication?

Domestication usually involves selection of genotypes that suit the intended human use. This will normally result in the domesticated plant population having only part of the variation that is to be found in wild populations. Subsequent selection following initial domestication only continues this trend to a narrower genetic variation (Figure 10.3). The domestication of sunflower illustrates this process. The sunflower (*Helianthus annuus*) was domesticated in the east central areas of North America. Molecular evidence suggests a single domestication, with a chloroplast genotype representing only 5 per cent of the wild populations in the region being the only one found in the domesticated population. Domestication and genetic selection following domestication each account for about half of the loss of diversity between wild sunflowers and modern cultivated sunflowers (Burke, 2008).

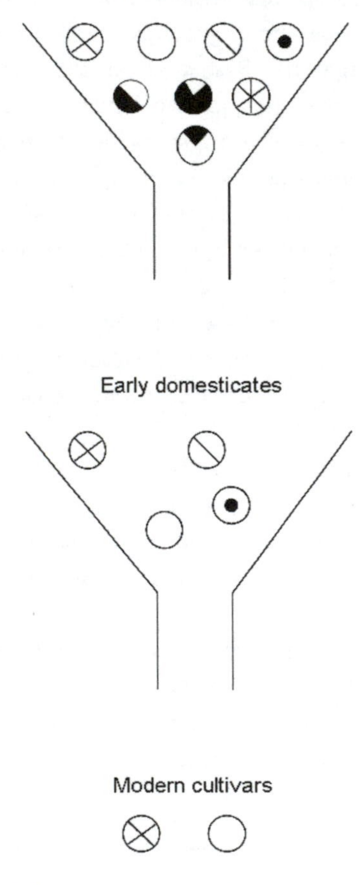

Figure 10.3 *Loss of genetic diversity in domestication*

In many cases the wild and domesticated species have not been completely isolated, and genes have continued to flow between wild and domesticated plants over a long period. The continuing interaction between wild and cultivated barley gene pools in the Fertile Crescent has already been mentioned in Chapter 2. The Azuki bean provides an excellent example of this process. Duncan Vaughan, from the National Institute of Agrobiological Sciences in Japan, conducts regular field trips to a site where a cultivated Adzuki bean (*Vigna angularis*, the source of red bean paste) is growing near populations of the wild plants. The cultivated beans are very distinct from the much smaller wild beans. On the edges of the cultivation, plants with seeds intermediate between the wild and cultivated forms can be found. DNA analysis has been used to confirm the evidence for gene flow between domesticated and wild plants at these sites.

In other cases the domesticated plants have become immediately reproductively isolated from their wild progenitors. Human removal of the plants to a new environment with no wild populations is a common mechanism for genetic isolation of domesticated plants. Plant domestication has also probably often involved human cultivation of genetic variants that suit human uses well, but may have lost reproductive capacity. The cultivation of polyploids (plants with multiple copies of their chromosomes) that in some cases have isolated themselves genetically from their diploid progenitors may be a recurring process.

Can we regain diversity from wild relatives?

Wild crop relatives have been widely used as a source of genes for improvement of crop plants. This is an especially important strategy for plant breeders when wild and domesticated populations have been genetically isolated by domestication events.

Knowledge of the relationships within plant groups provided by DNA analysis can be a very useful guide in the identification of sources of accessible wild genes for use in plant improvement. For example, the grape genus, *Vitis,* is restricted to temperate climates, but other members of the Vitaceae family are widespread in the tropics. DNA analysis has identified species currently classified within the *Cissus* genus that appear to be much more closely related to *Vitis* (Rossetto et al, 2002). Plants from these more tropical environments contain genes that could be useful in adapting grapes to more tropical climates.

Can we re-domesticate species for new uses?

Lima beans are an example of a single species that has apparently been independently domesticated to produce two different domesticated crops. This example supports the view that we can consider the option of domesticating species that have already been domesticated for food or other uses for new applications, selecting for different attributes and generating a crop with very different attributes. Lima beans have apparently been domesticated twice from *Phaseolus lunatus* (Motta, 2008). This is one of five *Phaseolus* species that have been domesticated. The lima beans, known as 'Big Lima', are of Andean origin, and were domesticated from wild populations on the western side of the Andes (Ecuador and northern Peru) and have seeds 10–14g. The small seeded (6–10g) 'Sieva' and 'Potato' lima beans were domesticated from wild populations in Meso-America. The squash, *Cucurbita pepo*, is another example of a species that has apparently been domesticated at least twice; once in south central Mexico and once in eastern North America and mid latitudes (Smith, 2006). These examples illustrate the potential to domesticate species more than once and to produce different domesticated outcomes.

Can we develop domesticated crops for new uses?

Flax (*Linum usitatissimum*) is a crop that was originally domesticated for oil around 10,000 years ago, but was later adapted to use for fibre production (Fu, 2008).

Options for domestication of new food crops

The potential to domesticate new food crops would be expected to be very limited because of the long period of time over which plants have been evaluated by humans in most parts of the world.

Exceptions might include species worthy of domestication in their own right that may have escaped domestication because there were not other species in the region that were suitable for domestication. Change from a hunting and gathering lifestyle to an agricultural one probably requires the availability of several species to make the agricultural option attractive. A good example may be found in the Macadamia. This Australian species was domesticated only very recently. The domesticated plants are derived from two wild species (*Macadamia tetraphyla* and *Macadamia integrifolia*) found in subtropical rainforests of central eastern Australia. The wild population of these species are now endangered. Wild trees were sent from Australia to Hawaii in the late 19th century, and these trees were used to establish Macadamia production there and were later re-imported to Australia to form the basis of a new industry from the 1960s.

Another area in which new possibilities for domestication may be found is when technology advances allow barriers to domestication to be overcome.

Figure 10.4 *Domestication in the Proteaceae*

The domestication of toxic plants as food crops by using modern genetics and chemical analysis methods to develop non-toxic cultivars illustrates this opportunity. The domestication of Canola from rapeseed may be considered such an example. A relative of the Macadamia, *Hicksbeachia pinnatifolia*, might prove to be another example (Figure 10.4). This species is a member of the Proteaceae family like the Macadamia and is found in rainforests in the southern parts of the range of the Macadamia species. The Hicksbeachia was discovered in the late 19th century and was described as edible in the records of early explorers. However, this now rare plant may not have been eaten by many humans. The author's experience of eating these fruits was one that suggests that under some conditions the nuts could be poisonous. Modern genetic and chemical analysis could be used to develop a more edible Hicksbeachia.

The Macadamia, originally a native of Australia, was largely domesticated in Hawaii. This cultivar, like most cultivated Macadamias, has a parentage involving contributions from both of the edible wild species, *Macadamia integrifolia* and *Macadamia tetraphyla*. *Hicksbeachia pinnatifolia* is a relative of Macadamia that has not been domesticated. This species has a range within the natural distribution of the two species of Macadamia that are the source of the cultivated Macadamia. The *Hicksbeachia pinnatifolia* is less likely to be domesticated because the nuts are not as attractive as a food and may also be toxic.

Many species are candidates for limited domestication. The Davidson's Plum – *Davidsonia pruriens* (Figure 10.5) – is found in the same areas as the species from the Proteaceae described above. The *Davidsonia* now include

Figure 10.5 *Davidson's Plum Davidsonia pruriens*

three taxa; *Davidsonia pruriens* from northern rainforests; *Davidsonia jerseyana* from the subtropical rainforests; and *Davidsonia johnsonii* from the same subtropical rainforests but without seeds. This later trait is of obvious interest for a cultivated fruit. Complete domestication may involve hybrid formation and exploitation of the seedless trait.

Davidsoniaceae is in limited cultivation for the edible fruits which are large and numerous, but not generally attractive to human tastes without the addition of significant amounts of sugar. The *Davidsonia* genus includes three taxa, two of which are rare rainforest species. The species are in limited domestication as a source of novel fruits. The reproductive biology of these species is poorly known, with *Davidsonia johnsonii* not known to have seeds.

The general conclusion that can be reached when looking at options for new food crops is that domesticating a new major crop with the value of rice, wheat, potato or maize is unlikely. However, regionally significant alternatives may be found and our reliance on such a small number of major food crops makes diversification highly desirable.

Options for domestication of new energy crops

The key genetic change that was selected in the domestication of the cereals was loss of shattering (as described in Chapter 2), so that the crop was not dispersed as it matured but stayed on the plant until harvested by humans. Domestication was also associated with strong selection for large seeds. Energy crops in which the total biomass is to be utilized for energy production will not require large seeds. Smaller seeds could be an advantage in production, allowing easier propagation of large populations of the crop. Non-shattering types will still be preferred, in sexually reproducing species, to allow for seed production. The genes determining both shattering and seed size have been identified and characterized in rice, and this knowledge should prove invaluable in rapidly modifying these traits in other plant species. Efforts to demonstrate the accelerated domestication of a wild Australian grass are in progress using targeted mutagenesis and a selection of natural variation in these genes.

The novel requirements of plants for energy makes it likely that plants better suited to this use have been overlooked by earlier domestication focused on food production. However, the potential to domesticate additional food crop plants is less clear. The selection of plants that have a high fibre or cell wall content for energy production will remove the potential for conflict with food use, as this type of biomass is not likely to be very edible or nutritious.

Domestication of new species starts with the selection of wild material, development of an agricultural production system, and then ongoing refinement of the crop by plant breeding and optimization of the production technology. Several examples of new energy cropping options will now be described.

Energy cane

Energy cane could build on sugarcane domestication to deliver greatly improved crops with energy and sugar as co-products. The utilization of the plant for energy could be more economic if sugar is extracted first and the remainder of the plant is used for energy. Sugarcane was domesticated by human selection of a fast growing plant with a high sugar content. *Saccharum officinarum* was domesticated (probably in New Guinea) as a sweet form of the wild *Saccharum robustum*. Genetic selection was for plants with a high sugar (sucrose) content in the stalk as a source of sweetness to satisfy human taste. Modern sugarcane cultivars have been developed by subsequent hybridization of the domesticated *S. officinarum* with wild *S. spontaneum* to improve growth rates without loss of sugar content. We may now wish to place more emphasis or total emphasis on the growth of biomass in the development of an energy cane. Modern cultivated sugarcane is a highly polyploidy species (it has many copies (often more than ten) of each chromosome). This high polyploidy makes breeding very difficult with the inheritance of many traits not behaving in the simple way that they do in diploid (only two copies of each chromosome as in humans) plants. In this extreme case the domestication of a plant that was less complex genetically may be a worthy objective. This would allow more rapid breeding. Sugarcane is propagated vegetatively (using cuttings) and has gone through relatively few generations of sexual reproduction since domestication. Modern cultivars are probably only a few sexual generations from wild plants. The related grass, Miscanthus, is being widely evaluated as a potential energy crop. Crosses between Miscanthus and sugarcane have been produced as additional options for energy and sugar production.

Energy grain

Sorghum could be de-designed as a multipurpose food/feed and energy crop. The entire plant could be utilized with the grain contributing either starch to biofuel production and protein to feed uses or to novel food products, with the remainder of the plant being utilized as a source of energy. Domestication of sorghum has focused on grain traits as this is the part of the plant eaten by humans or animals. The vegetative growth of the plant would become the primary target for breeding for use as a biomass source for biofuels. Recent efforts have focused on sweet sorghum as an option that allows first-generation biofuel production from the grain and vegetative parts of the plant. The focus should shift to the potential of sorghum as second-generation biofuel crop. Total biomass production rather than grain or sucrose content is the target for this research. Several species that have been identified as close relatives of cultivated sorghum by DNA analysis can now be included in the germplasm pool for use in sorghum breeding. These may increase the range of traits that are available in sorghum.

Energy tuber

Tuber crops such as potato, sweet potato and cassava may contribute to energy production. They have potential for current use, especially use of waste components of these crops, in first-generation fuel production. The role of these high-value food crops in second-generation biofuel production is less clear.

Energy grass

New grass crops could be domesticated as dedicated energy crops. Total biomass production with minimal inputs would be the main selection criteria. Biomass composition could also be optimized for specific biofuel conversion technologies. Domestication of grasses as energy crops requires some of the same steps as the domestication of the cereals as food crops. For example, a reduction in shattering is still required to allow harvesting of the seed for propagation of the crop rather than harvest of seed as food. However, large seed size beyond that necessary to ensure good seedling establishment may not be an advantage. *Miscanthus* species and switchgrass (*Panicum virgatum*) are good examples of grasses being evaluated as new energy crops (Vermerris, 2008b).

Energy wood

Novel woody biomass crops such as Eucalypts could be developed. Poplar (*Populus* spp.), willow (*Salix* spp.) and pines (such as *Pinus elliottii*) are also important options. Options to maximize the sustainable harvest of biomass on an annual basis would require the determination of the optimal harvest frequency (e.g. every year, two years or ten years) to minimize the costs of harvesting and deliver the maximum yield of biomass per year. Breeding for this production system would require the selection of species with a long life between replanting combined with maximal growth rates. Novel changes in biomass composition could be introduced to enhance the value of this biomass in biofuel production. A very large number of species adapted to a wide range of environments are available to select as energy crops.

Energy legume

A legume crop, or at least one that results in increased incorporation of nitrogen into the soil, may be an important option in many regions. These species reduce the reliance on nitrogenous fertilizers produced using energy-intensive processes and including the use of fossil fuels. Pongamia is a legume that has been developed as a first-generation, oil producing, energy crop. The development of second-generation energy crops from legumes is now needed. These species may have value as a crop in their own right. However, their main use may be as a rotational crop delivering advantages in nutrition to the crops that are planted following them in the same field. This is important for both food and energy production. A legume with food or energy uses (or both) could rotate with a non-legume with either food or energy uses (or both). The economics of the whole system need to be considered. Ideally, the value of each crop is maximized in addition to considering the value of each crop in the rotation.

Urgent application of these strategies is needed to generate the new crops required to satisfy human energy requirements.

Novel tools for domestication

An improving understanding of the processes of domestication has come from the availability of genetic analysis tools. Genetic relationships between wild and domesticated populations have been used to define the processes that resulted in domestication. The analysis of DNA sequences from domesticated plants has defined some of the exact changes required to adapt a plant to domesticated production. For example, the genes controlling shatter in cereals are being defined. Manipulation of these genes could rapidly alter this key domestication trait. Similar approaches might be employed to target other important domestication traits. New and emerging tools are greatly increasing our capacity to analyse large amounts of DNA from large numbers of individuals in increasingly efficient and low-cost protocols. These scientific advances offer the potential to greatly accelerate our ability to domesticate new species. We can expect rapid growth in our knowledge of domestication genes and processes in the next few years, as these new technologies are more widely applied. The application of genomic approaches is likely to further accelerate research in the accelerated domestication of new plant species. Genomics involves the study of all (or most) of the genes in an organism as opposed to conventional genetic approaches that consider the role of individual genes. The positive impact of this technology on the breeding of food and energy crops has been discussed in Chapters 2 and 5. Advances in DNA sequencing technology (see Chapter 5) since about 2005 have been very rapid, making this technology much more widely available and feasible for this type of application. DNA banks are improving access to biodiversity for this type of research. This technology is continuing to be applied widely to food and forest species, and can now be used to accelerate the domestication of energy and new food crops to rapidly catch up on much of the 10,000 years or more that the traditional food crops have. Improvements in biomass composition and specific traits will become increasingly feasible as the science improves. Major advances in total plant biomass yield may be slower and require much more field evaluation and a greater understanding of the performance of plants in each local target environment.

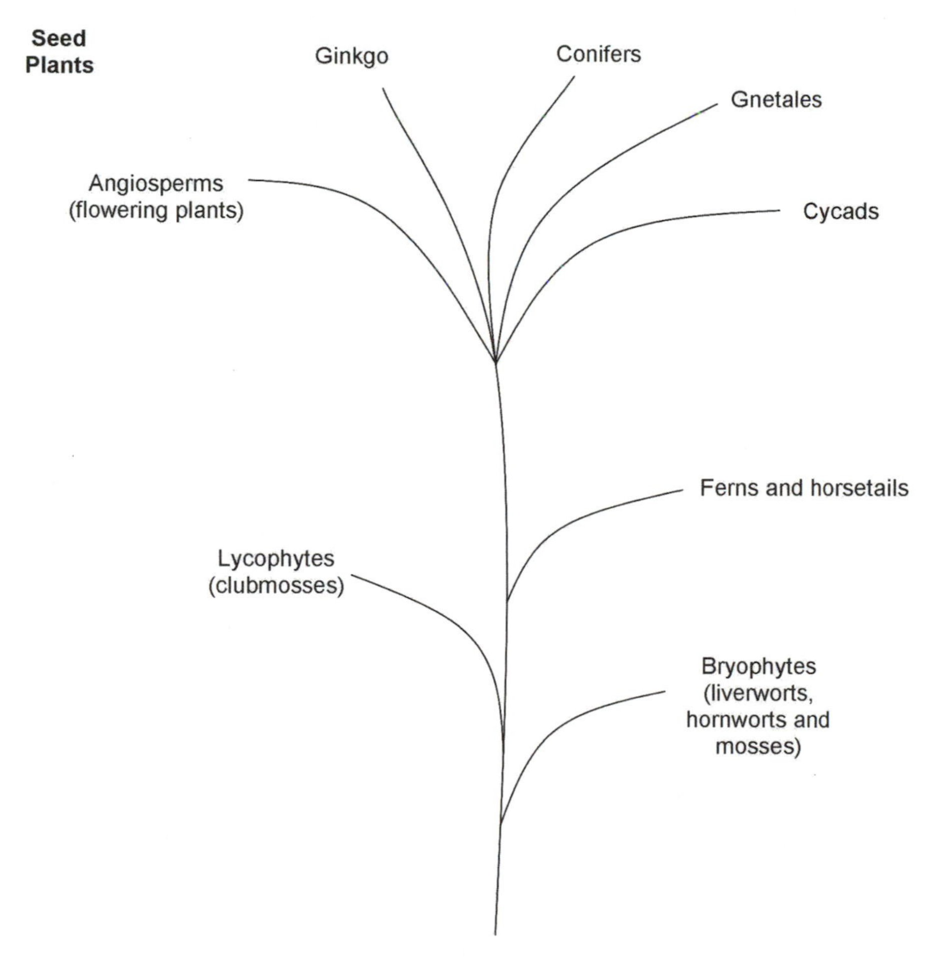

Figure 10.6 *Relationships between higher plants*

Humans have domesticated many plants for food, but it remains difficult to identify a plant species that has yet been domesticated for energy. Some plants may be considered to be currently in the early stages of domestication as energy plants. Some of these may continue to full domestication. However, it is likely that new species not yet even under consideration may be rapidly domesticated and become our dominant energy crops. All of the plants should be considered options (Figure 10.6). However, the angiosperms or flowering plants are the dominant large plant group and represent the main option.

A systematic analysis of the families of angiosperm and gymnosperm families and their food, energy, or other current or potential use, is provided in Henry (2010).

Chapter 11 will look at options for the future use of plant resources for food, fuel or conservation.

11

Options for the Future

As potentially environmentally sustainable commodities, the enthusiasm for plant-derived products is understandable. In principle, a deeper understanding of plants and other living systems could allow us to better manage the earth's resources for both environmental and economic ends. Demand for land, water and biomass resources is intensifying with consequences that are being felt by all. If current developments are anything to go by, the politics of plants will quickly become increasingly complicated.

Frow et al (2009)

The demands of human populations for food and possibly energy from agriculture are likely to be sustained, over the next few decades at least. This will be driven by continued population growth and increased food consumption associated with economic growth in developing countries, and increased use of bioenergy as an alternative to fossil fuels in an attempt to minimize the impact of climate change. Conserving biodiversity will continue to be a challenge, while pressure to commit more land to agriculture will remain an enduring reality. The greatest uncertainty is probably the extent to which climate change will exacerbate these problems by reducing agricultural productivity, as a result increasing the demand for land for agriculture; and by directly driving species to extinction. This chapter will define the issues and possible solutions. Key aspects of these challenges to be covered include the need to focus on strategies to minimize food energy competition both at the biomass and land-use levels. The areas in which science can contribute need to be systematically evaluated.

Historically food production has increased to meet or exceed population growth and associated demand for food. In the second half of the 20th century this was achieved in the 'Green Revolution' that combined new higher yielding genotypes with the use of increased inputs such as fertilizers, but with very little change in the area being cultivated. In the last decade the view has emerged that this technology has reached its limits with a slowing in the rate of growth of food production (Fedoroff and Brown, 2004). Some people argue that scientists are to blame for the growth in human populations because they developed more efficient agriculture that has kept pace so far with the growth in populations. The alternative to not developing the capacity to feed a growing population would be limiting growth by starvation. Surely the ethics of scientists and the community at large must be to support human life, while highlighting the strain population growth places on global sustainability? However, the 'Green Revolution' has not been without significant problems. A consistent criticism is that the technology has made farmers in developing countries more dependent on high cost nutrient inputs. The reality is that the new plant cultivars tend to yield more regardless of fertilizer use, but do respond very well when additional nutrients are used. Loss of biodiversity in agriculture due to the dominance of high-yielding crop cultivars that encourage the growth of monocultures is another continuing concern. Plant breeders are now able to directly tackle this problem by deliberately introducing more diverse material into their programmes. Modern DNA analysis tools assist in maximizing diversity in cultivated plants by allowing deliberate selection of new cultivars or parental lines for use in breeding that are as genetically different as possible from other cultivars in cultivation. This needs to continue as a very active process to balance the relentless pressure on agriculture to maximize performance by selecting the very best performing genes or genotypes. These deficiencies do not detract from the key role this technology has played in feeding the human population and protecting biodiversity by minimizing the footprint of agriculture, but they do dictate a need for continued efforts to address the unintended problems that result.

Climate change and the costs and limits of fossil fuels have provided a strong incentive to explore the potential for efficient energy crops. However, competition for land with food crops and biodiversity conservation are potentially serious negative impacts of biofuel production. This suggests the urgent need to focus biofuel crop production on species that are highly efficient and do not compete directly with food crops for land or water. The ideal species for these applications have not been identified for most production environments worldwide. Time has probably been wasted on attempts to adapt or use food crops rather than undertake the testing required to reliably identify and in many cases domesticate new species as biofuel crops from the many plant species available (see Chapter 11). The long-term priority for biomass production will probably be to replace fossil fuels in applications other than transport fuels. Biofuels provide a partial or possibly a complete solution for transport in the short term (next few decades). However, they will only make a net positive contribution if their production and use is managed better than it is at present. New technologies will be needed to supply energy for transport in the long term.

Scenarios for the future requiring differing degrees of success in these areas can be defined. It is worth thinking about options over a period corresponding to the current life expectancy of someone born today in a developed country. The lifespan of this person may depend on how the different scenarios play out, with the possibility of it being much longer (due to advances in medical research) or much shorter (due to environmental decline) than current expectations. The scenarios explored here are to the 2085 time point based upon the calculations of Cline (2007), as introduced in Chapter 6. These estimates are based upon two population projections for 2085; a low population projection of 10.5 billion and a high (worst case) projection of 14.7 billion.

These calculations suggest a 2.66-fold (low population) and a 3.72-fold (high population) growth in food demand by 2085 compared with a starting point of 2005. Growth in food supply was estimated to be 2.44-fold in these scenarios. This immediately suggests we should be using all our efforts to aim for the lower population outcome. One complication is that an outcome of lowering population growth may be associated with a further boost to per capita consumption (already as big a factor as population growth in determining food demand). This would reduce the benefits of population control. Affluence is generally associated with lower population growth rates, as economic concerns rather than survival determine human reproductive behaviour.

We can now consider the impact of our approach to biofuel production on these scenarios in a low biofuels future and a high biofuels future.

Low biofuels future (next two decades)

This low biofuel option avoids tackling replacement of fossil fuels and has a great risk of resulting in continued high levels of greenhouse gas accumulation and associated climate change, with serious implications for ongoing food security and biodiversity conservation. The low biofuel option only looks good for climate change mitigation if it is achieved by making biofuels redundant, by developing new alternative energy technology that is carbon neutral and can replace most uses of fossil fuels. This is relatively unlikely in the next two decades but should remain a longer-term objective. The benefits of this option may include reduced competition with food production and the associated resulting lower pressure on biodiversity. This needs to be balanced, however, against the high risks of climate change, and a resulting loss of agricultural productivity and direct loss of biodiversity. The high cost of fossil fuels may also drive up food production costs in this scenario.

High biofuels future (next two decades)

Second- or later-generation biofuels (defined in Chapter 5) should allow a reduction in greenhouse gases in the short to medium term, allowing time for alternative technologies to be developed to replace carbon-based fuels in the longer term. This option only looks good if we are able to develop highly efficient second-generation fuel technology. Current first generation technology is too inefficient, resulting in a much larger environmental footprint for biofuel production.

The attractiveness of this option relative to the low biofuels option suggests that we should place great emphasis on achieving efficient second-generation technology quickly. Despite the greatly improved efficiencies promised by second-generation technologies, the potential for food and biofuel production to compete for land, water and other resources needs to be managed carefully and actively, as does competition with land for biodiversity conservation.

The choice between these two options becomes one related to the relative difficulty and likelihood of success in developing these technologies (new carbon neutral energy or second-generation biofuels). In reality we need both technologies as soon as possible. The high biofuels option is only a partial solution to our long-term energy needs. If we have sustainable cellulosic biofuel we will still need other technologies to satisfy our total energy needs.

Scenarios for the future

Biofuel production is likely to be an essential component of delivering a sustainable future. We need to begin to reduce CO_2 emissions. Large-scale adoption of biomass-based biofuels has been estimated as being capable of replacing 30 per cent of US oil production by 2030. This would require 1 billion tonnes of biomass and it has been suggested that this is feasible (US, DOE, 2005). This could be combined with the replacement of another 20 per cent of oil consumption with solar and wind energy on the same timescale. Many other countries have a much greater capacity to produce biofuels relative to their energy use. This type of scenario would project the possibility of CO_2 emissions peaking by 2050 and make significant progress towards an end to substantial use of fossil fuels by the end of the century (Ahmann and Dorgan, 2007). While this is a relatively easy path to follow, many suggest we need to act much more aggressively than this. Larger volumes of biofuel production to replace most or all of fossil fuel use on a shorter timescale would require the production of dedicated biofuel crops.

Lessons from the past

Can we learn from previous human experiences and avoid a disaster in the supply of food and energy, but at the same time conserve biodiversity? Diamond (2005b) has examined a number of examples of dramatic failures of human societies in his book, *Collapse*. In a similar way climate change may result in food supply failing to satisfy the needs of large human populations.

Box 11.1 *The case of the Mayan society*

The downfall of the Mayan society provides lessons that have many parallels with the current situation. The ruins of this society in Mexico (Figure 11.1), much still overgrown by forest, should be a reminder of what appears to have happened when a society was unable to feed itself. The Mayan society collapsed rapidly around 1000 years ago. The society was highly dependent on maize and had limited water storage facilities for crop irrigation. This region has a highly variable rainfall, making the society vulnerable to a series of low rainfall years. Climate change – even short-term fluctuations in the climate – may have a catastrophic impact on a society that has a large population with very limited amounts of stored food and is reliant on a very limited number of crop plants for agricultural production. This parallels human society today with our great dependence on a few crop species, the products of which we have little of in storage at any time, with a large and growing population facing the prospect of climate change.

Figure 11.1 *Mayan ruins on the Yucatan Peninsula, Mexico. These societies were based upon maize. The rock carvings (inset) depict maize grains and maize leaves*

Was Ehrlich right?

Paul Ehrlich predicted that the world would face widespread starvation if population was not controlled in his work *The Population Bomb*, published in 1971. This work and *The Limits to Growth* (Meadows et al, 1972), and its follow-up volume *The Limits to Growth: The 30-Year Update* (Meadows et al, 2004), shaped the views of many and can be linked to the awakening of wider environmental awareness of resource constraints in society. The predictions of Ehrlich's book proved inaccurate in so much as the impact of technology allowing us to feed a growing population was underestimated. But effectively we may have just delayed the inevitable. The alternative is to assume that technological advances will continue to allow us to avoid disaster. It is important that if this is our assumption then we acknowledge that this is the basis on which we go forward and consider the consequences of this strategy. If we bet the future on technology advances we need to make sure we resource the science and technology necessary to deliver these required outcomes. Public policies generally fail to acknowledge this link and often support continued expansion of human activities, while withdrawing or not supplying the resources required to develop the technology to deal with the consequences.

Does new technology offer solutions?

The key areas of technology that are essential to our ability to sustainably produce food and energy for the human population are in plant production (agriculture and forestry), in the processing of crops to food (food processing), and in producing energy from biomass (conversion technologies).

Crop production

The technology inputs in food crop production can be divided into the genetic component and the management component. The genetic component is the contribution of the plant breeder to selecting and breeding a cultivar with high yield, and with the desired functional and nutritional properties. The management component (agronomy and farming systems) includes the entire process of where and how we cultivate the crop.

Relative contributions of plant breeding and better crop production systems

Both genetics and management have been major components of the improvements in crop productivity in the past, and both must be pursued aggressively if we are to continue to match food production to human demand. The relative contributions of these inputs have varied in different regions and for different crops, but are probably each responsible for about half of the improvements we have made and are likely to continue to make. Genetics and management have made synergistic contributions to achieving our food production needs.

Plant breeders argue that the advantage of the genetic contribution is the ease of adoption. A new cultivar can simply be supplied to a farmer as seed, and all the benefits in performance and environmental adaption built into the cultivar in the research and development are transferred to the end user. Many technologies, including the production of hybrid cultivars, continue to offer the potential to greatly increase the productivity of crop production systems. Wide hybrids (hybrids with genetically diverse parents) result in hybrid vigour (heterosis) that gives much greater productivity in many plant systems. Maize hybrids are probably the best known example of this technology, with great advances in productivity being achieved in corn production using hybrids. Hybrid Eucalypts with greatly improved performance have been recently developed as ornamentals (Figure 11.2), but may also become a major opportunity for energy crops. Hybrid tropical pines have been developed and successfully deployed for plantation forestry. For example, timber for the frames of houses in Queensland has been largely sourced from

Figure 11.2 *Hybrid Eucalypts – Hybrid vigour. This inter-specific Eucalypt hybrid was produced as an ornamental, and shows growth and floral characteristics superior to either of the two parental species*

plantations of tropical pine trees (*Pinus*), with the Slash pine (*Pinus elliotii* var. *elliotii*) and the Caribbean pine (*Pinus caribaea* var. *hondurensis*) being grown on different sites to satisfy this requirement. The Slash pine is better adapted to lower, wetter areas and was favoured for planting in these environments. Better drained areas were planted with Caribbean pine. A hybrid between the two has been found to be broadly adapted and suited to planting across the entire area of forest plantations, and has been replacing the pure species in recent plantings. This is a good example of how breeding technology has impacted to improve the management efficiency of forest tree production. More widespread application of hybrid technology can be expected to continue to provide an important tool in meeting the critical demand for improvement of the productivity of food, forest and energy crops.

Plant-breeding technologies, especially the efficiency and accuracy of genetic screening, have been enhanced by the development of molecular techniques for the selection of plants with desirable production traits. Advances in genomics promise to greatly increase the effectiveness of this technology in the future as we learn more about the genetic basis of many key plant traits. Automation of these molecular screening techniques continues to advance, suggesting that the widespread and cost-effective application of these technologies will play a central role in accelerating the genetic improvement of plants for all end uses.

Box 11.2 *Advances in DNA fingerprinting techniques for use in plant improvement*

The future will see direct analysis of plant DNA for the traits of importance to humans, rather than the indirect approaches of recent decades that have required analysis of DNA for genetic markers that are not always perfectly linked to the trait of interest. DNA analysis in plants has had a wide range of uses.

DNA-based plant identification has application in:

- the protection of intellectual property rights associated with plant cultivars;
- forensic applications in the analysis of crime scenes;
- determining the identity and purity of seed lots in commercial trade and production;
- determining the cultivar of an agricultural species that is being traded to ensure appropriate contract compliance;
- food processing to ensure the processing method is optimized for the cultivar;
- analysis of the composition of competitors' food products.

These same DNA fingerprinting techniques are useful in selection and breeding of new plant cultivars with desirable production traits (see Chapters 2 and 5). The technology used has been generally based upon the approaches used in the DNA fingerprinting of humans, largely for health (genetic disease analysis), paternity and criminal identification. The technologies used have evolved over the last decades (Henry, 2001; Henry, 2008) to become more reliable, automated and low cost. The technology has now advanced to the stage that the largest component of the cost of most analyses is the cost of collecting and handling the sample. The utility of this technology will be greatly enhanced by the increased knowledge of plant gene sequence flowing from recent technology developments.

Advances in the genomics (study of all of the genes) in plant species (Box 11.3) have underpinned the rapid growth in understanding of gene function necessary to apply selection at the DNA level in plant breeding. The analysis of the sequences of plant genomes is a great platform of information on which this technology is now developing rapidly.

Box 11.3 *Plant genomics*

Plant genomics provides enabling technologies for developing the plants of the future and for monitoring the status of plant biodiversity. New and emerging technologies are accelerating the rate of discoveries in plant genomics.

Key developments include:

- rapidly increasing capacity to determine the DNA sequence (genetic code) of any plant;
- growth in functional genomics (understanding how plant genes function);
- improved understanding of the genetic basis of nutritional and functional characteristics of foods;
- insights emerging into how plants may adapt or be adapted to climate change;
- greatly improved ability to efficiently select plants with desirable characteristics in plant breeding;
- related improvements in DNA fingerprinting for identification of plant cultivars in production and food processing and distribution;
- emerging tools for objective DNA-based biodiversity assessment and monitoring;

- improved understanding of evolutionary and adaptive processes in plants;
- tools for determining relationships between individuals in critically endangered plants and managing populations for long-term survival and diversity.

Improved ways to grow the crop and manage the overall farming operation may offer very large opportunities for improved productivity and sustainability, as well as biodiversity conservation. However, the impact of research and development aiming to deliver these outcomes will depend upon the extent to which new management practices are adopted by farmers. The communication of the technology may be frustrated by the necessarily conservative nature of farming operations. Many farmers even in the developed world tend to farm the way their parents did, resisting new approaches. Demonstrations of the benefits of the new approaches on the farmer's own land or on the farm of a neighbour is a strategy that is often necessary to convince farmers to change long-established practices.

We need both strategies (genetics and management) to be able to satisfy future food and energy needs. New plant cultivars need to be produced using new management strategies to allow the necessary increases in productivity and sustainability. The synergies of these interdependent developments are essential to satisfying demands for agricultural production.

Food processing

Significant gains in food supply may be achieved by improving the efficiency of food processing. Increasing the yield of flour from wheat is a good example of this approach. Most wheat (in excess of 500 million tonnes per year) is processed by milling to produce flour. The flour accounts for the bulk of the grain, but significant amounts are separated as bran that despite a high nutritional value is not consumed in large quantities by humans. Genetic improvement of wheat to produce cultivars that yield more flour on milling, or improvements of the milling process to recover more flour, are options for increasing the supply of food from wheat. These processing advances provide more food without the need for more land, water or other farming inputs. Because of these advantages these approaches should be given high priority. Protection of plants against post-harvest spoilage may also make a significant contribution to food supply.

Box 11.4 *Research targeting better health and functionality in foods*

Much research effort is being devoted to developing foods that have the potential to contribute to improved human health and provide improvements in the yield and quality of processed foods. Two examples of programmes aimed at delivering these outcomes are the Healthgrain programme in Europe and the Grain Foods Cooperative Research Centre in Australia.

Healthgrains

Healthgrains (www.healthgrain.org) is supported by the European Union are with contributions from many food companies and research organizations in Europe. Grain quality attributes such as fibre content, important for human health, are major targets of this research. The project 'Improving health by exploiting bioactivity of European grains', or Healthgrains, aims to deliver whole grain products or grain fractions that contribute to improved human health.

Grain Foods CRC

The Grain Foods Cooperative Research Centre (www.grainfoodscrc.com.au) is supported by the Australian government and food companies and research agencies. Research aims to improve the efficiency of recovery of quality foods from grain and to deliver products with health and processing advantages. Examples of research outcomes include genetic discoveries that will allow the production of grains with enhanced folate content and improved flavour. Improvements in both the plant cultivars and the processing techniques are combined to deliver these outcomes.

Another good example of the type of innovation in this area is the development of low-allergy peanuts (Chu et al, 2008). Peanuts are a high-value food, in both developing and developed countries, that can nevertheless cause a serious and fatal allergy in a small portion of the population. Increasing understanding of the proteins in peanuts that are responsible for the allergic reaction in humans has defined targets for research, aiming to develop peanuts that will not be allergenic or at least will be low in allergens. The specific proteins in the peanut that cause the allergy are now known and the genes expressing these proteins can become targets for efforts to turn them off by genetic manipulation, or by mutation or even selection within natural variation. More widespread application of this type of approach may assist the diversification of human food sources by allowing food use of plants that are currently not considered edible because of toxins or other health risks.

Conversion to energy

Improved technologies for conversion of crop biomass to energy also have a great contribution to make. Developing the ability to convert all of the carbon (or at least all of the carbohydrate, structural and non-structural) to energy is an essential step in achieving the efficiencies required to make this technology attractive from an economic and environmental perspective. This advance would facilitate the application of strategies to manage the efficient utilization of total biomass for food and energy production, and allow the capture of all current waste streams from agricultural and forest production and processing. Multipurpose crops might become a major option with the whole plant being used for food and energy, as discussed in earlier chapters.

Genomics is promising new insights into how to improve crops based upon identification of the genes controlling the composition of plant biomass. The structure of cellulose, non-cellulosic polysaccharides, lignin and the linkages between these components are all targets for development of novel plant biomass that facilitates efficient conversion to biofuel.

A significant example of the type of innovation that the future might bring is provided by the recent report of new insights into the control of cell wall formation in plants (Held et al, 2008). This research has identified genetic mechanisms that influence the transition from primary to secondary cell wall synthesis in plants. The primary cell wall is formed while the cell is still enlarging. Mature cells may then form a secondary cell wall after cell growth has been completed. The composition and extent of the secondary cell-wall is very dependent upon the cell type. Small interfering RNAs have been shown to influence the expression of the genes encoding the major enzymes of cellulose and non-cellulosic glucan (glucose polymer) biosynthesis in barley. This research suggests options for manipulation of the biomass value of grasses and possibly all other plants.

Even more novel organisms might be engineered in the future to facilitate conversion of biomass to biofuels, or even direct biofuel production. The production of synthetic genomes has been proposed by Craig Venter as a strategy for biofuel production and the process of synthetic genome construction has already been demonstrated (Gibson et al, 2008). This technology creates complete micro-organisms to human designs by chemical synthesis of DNA sequences encoding the functions required in the organism. This could be used to generate organisms that produce biofuel molecules directly from carbon dioxide, using light from the Sun as a source of energy. Alternatively, organisms could be engineered to convert carbon capture by plants into biofuels.

Organisms that are found in extreme environments may provide some of the solutions to biofuel production. These organisms are found in environments that are extreme in temperature or chemical composition. High-temperature volcanic sites or soils that are very acid or alkaline may provide these unique organisms. These organisms or enzymes isolated from these organisms may have properties that allow their use in novel industrial processes.

The production of higher value fuels may be achieved by using organisms that are able to produce these molecules by fermentation. For example, changing from ethanol to butanol may simply require the use of a different micro-organism for fermentation.

New technologies for biodiesel production are also likely. The use of supercritical fluid extraction may replace the use of organic solvents to provide more efficient and more environmentally desirable processes for oil extraction from plants. More efficient methods for production of biodiesel from cellulosic biomass are also required.

Integrated food and energy production systems

Multipurpose crops (in which part of the plant is eaten and part used for energy), or crops that can be grown together in the same field to address different end uses, are important options. Increasingly, food processors are examining all waste streams as options for use in energy production. It is likely that these options will be among the first to be exploited because the cost of harvest and transport of the biomass to a central location has already been invested.

Contribution of plants to carbon storage

The use of plants to capture carbon may also become important. Changing land use towards the production of large plants such as trees may increase the amount of biomass in plants at any one time. This will increase the amount of carbon in the standing biomass. Crop residues contribute carbon to the soil, which has an important role in retaining soil nutrient status and allowing sustainable production. Some of the carbon from crop residues is in forms that may persist for long periods in the soil. This allows carbon to be stored or sequestered and may assist in reduction of the amount of CO_2 in the atmosphere. Small amounts of carbon may be trapped inside cells – phytoliths – with walls rich in silica that protect the contents long term. The burning of plant residues may generate biochar that also has a relatively long life in the soil. The selection of crops may be influenced by these environmental values. Plant breeding may be used to develop crops with greater potential to contribute to carbon storage in these ways.

What type of future are we creating?

If we agree that humans can continue to expand their population indefinitely using technology to deal with the problems as they arise, then we need to understand that that is our strategy and accept its consequences for the type of future world we are creating. We cannot argue that all humans are not entitled to the same environmental footprint. However, we need to be aware of the huge impact of the world population moving to the level of resource consumption of those in developed countries.

Human evolution is increasingly locking us into a future environment that is of human creation. As we occupy more of the space on Earth and continue to reduce the numbers of other species and their contribution to the biosphere, we become more reliant on our science and technology for our survival. The ultimate step is probably for the human species to develop the capability to survive beyond this planet. This may become more than an option, rather a necessity, if we continue to overrun the biological systems on Earth. In some ways continuing on Earth may become almost the same option if we need to create an artificial environment on Earth for survival. This is a dangerous path, especially if followed at the rate we are at present. Can we be sure we can continue to find the technical fixes in time as we exhaust the capacity of the Earth to support our needs for food, energy and a survivable environment? A slower rate of advancing down this path would help. Slowing these processes down to give the human species a better chance of surviving the future implies greater efforts to retain the biological resources of Earth as long as possible. Some actions we can take to contribute to a more sustainable future are listed in the next section.

The development of the Internet has undoubtedly been a major advance in human communication. The rapid exchange of information creates the potential for a much more global approach to our problems and much greater international understanding and cooperation. Humans are also becoming more disconnected from the biological systems on which they depend. We are already spending less time relating to nature as access declines and more to systems created by humans such as the Internet. Environmental education needs to adjust to this reality and aim to ensure we do not lose sight of our origins.

Conscious selection of the future

Growth of human populations has largely driven our expansion and the consequences of this growth have been dealt with as they arise. Should we just let events decide themselves? Humans need to consider the balance they want in their future. Emphasis to date has been on the quantity of food and energy consumed per person as measures of our quality of life. Energy allows travel, thus enriching human lives by exposure to diversity of human cultures and to biodiversity. However, travel and communication is also accelerating the development of a more homogeneous world, as well as contributing to climate change, and we face the prospect of a continuing loss of biodiversity. An alternative emphasis would preserve more of the diversity of biological experiences available to humans. Quality of life can also be derived from the diversity of life forms we have the opportunity to interact with and the diversity of foods (based upon biodiversity) that we have to eat.

A future in which we have the greatest possible diversity of life forms seems to be more attractive than one in which we simply have enough food and energy to survive.

We will now consider the implications of these two contrasting futures.

Low biodiversity future

The extension of many current trends will see a low biodiversity future. In this scenario species extinctions will continue, probably at an accelerating pace. The outcome will be that human life will be dependent on a narrower biodiversity base. This will carry significant risks. It will directly limit our ability to adapt food production to environmental change. It will mean we rely on fewer species to maintain the biological cycles that are essential to maintaining air, water and soil, indispensable to a life based upon biological inputs. The science and technology required to continue life without these resources is likely to take a very long time to develop if it proves possible. We are on a path that will increasingly lock us into a future that anticipates such dramatic and potentially unattainable developments.

High biodiversity future

A high biodiversity future is a much safer option. It gives us more time to develop the science and technology required to cope with our future. It provides a reservoir of biological resources to support adaptation to a changing environment. It will ensure evolutionary processes can flourish, delivering the biological resilience to ensure life continues on Earth long term. This outcome requires that we focus on the issues of sustainability discussed in this book. Recommended actions that will contribute to ensuring this are given in the following section.

Regional or local solutions

The outcomes of efforts to balance the competing needs for food, energy and conservation of biodiversity at a global level will be the sum of all local or regional outcomes. The environment differs from place to place in many ways that influence the choices that are available. Local analysis is required to identify the best local solution. Agriculture in any given region has developed in response to the experiences of local farmers in working with the soil and local climate to deliver crops that are profitable. Subtle differences in day length, temperature and rainfall (total amount and distribution over the year) are among very many environmental variables that determine the choice of crop species and cultivar, the planting time and other management requirements. Over time sustainability becomes more important as farmers seek to continue repeating their crop success each year by adopting more sustainable techniques. Many agricultural scientists report that their experience of working with farmers in many different regions has been that they are interested in the sustainability of their businesses. For them environmental sustainability is economic sustainability. The converse is also true; without economic sustainability they cannot afford to consider environmental sustainability. The difference is only in the timescale, especially in variable environments when the production of a good year needs to cover potential losses in less favourable years in the cycle. The global balance is also important to ensure we meet the needs of the planet at a higher level. Unless this is monitored and actively managed we run the risk that outcomes at a global level will not be balanced. We need to continue to work at both a local and global level. Agricultural research and development has declined in developing countries in recent years. Increased investment in this area is essential if we are to maintain food security and improve sustainability (Von Braun, 2008).

The future of human life on Earth depends on a globally sustainable environment that remains suitable for human survival. The management of the future will require complex decisions and actions. More biologically diverse futures will be more stable and more biologically secure. The quality of human life will also be enriched if we can retain significant biodiversity. A scientifically literate society is the base on which this future depends. If we fail to provide rigorous scientific training to large numbers of people we will lose the capacity to respond to future needs. Indeed, this may be the greatest threat to a sustainable future. We need many technologies and many solutions both local and global, and to achieve this we need many active minds seeking sustainable options.

Recommended actions

The discussion in this book has directly or indirectly defined actions that can be taken to improve or diversify future options. Twenty-five key technical and policy objectives (not in priority order) in support of a sustainable food, energy and biodiversity future for humans on earth are:

1 *Increase efforts to conserve genetic resources of the major food crop species on which we all depend for food (especially those such as wheat, rice, potato and maize)*
The critical dependence of humans on these few species for food requires that genetic resources are available to allow adaptation to changes in climate and the outbreak of new strains of major plant diseases. The loss of any one of the major species would put great pressure on food supplies. Although the collections of germplasm for these species seem large, much of the store material is redundant (many copies of the same genotype stored), and as our major sources of food they require continued efforts to conserve all available diversity. Conservation of diversity of these species needs to be balanced against the need to diversify support for food crop species to the many minor food crops. However, the overwhelming importance of the major species in global food supply cannot be ignored.

2 *Identify and protect more areas with high biodiversity value in national parks and reserves*
National parks and reserves with resources to ensure they are protected will be very important in the conservation of biodiversity on an increasingly crowded planet. The conservation of plant biodiversity outside protected areas needs to also be addressed. However, the conservation of plants in protected areas is essential because many species are found in very restricted habitats that require special protection to give these species any chance of survival.

3 *Improve the palatability of more stress-tolerant food crops (e.g. barley and sorghum) that are better able to cope with climate change and develop food technology so that these species can contribute new options to meet the demands of food supply*
Diversification of the species used as staple crops will expand the range of environments that are able to make a major contribution to human food needs. Making productive species more palatable as foods needs to complement the more common strategy of increasing the productivity of attractive foods. Barley and sorghum are especially important targets as they are well adapted to marginal environments and close to being acceptable as human foods. Most other species are

either not suitable for very widespread production or a long way from being highly palatable to humans.

4 *Develop technologies to achieve maximum recovery of food from harvested crops and reduce post-harvest losses of foods in transport and storage*
Improvements in the recovery of food from harvested crops make an especially valuable contribution since the water and land resources required have already been consumed in the production of the crop. Most other options for producing more food will put more pressure on these resources.

5 *Develop new second-generation energy crops by adapting current crop species (especially grasses and trees that can deliver high biomass quantities from less favourable environments (e.g. disturbed areas or disused agricultural areas of low biodiversity))*
Energy production that has a minimal potential to compete with food production or biodiversity conservation is the outcome this will support.

6 *Perfect ligno-cellulosic (second-generation) conversion technologies so that they are efficient and environmentally sound*
This will provide the maximum potential energy and greenhouse gas reductions that biofuels offer.

7 *Ensure public policies on agriculture, food and energy production encourage sustainable technologies that are compatible with outcomes that include biodiversity conservation as a key objective*
Biodiversity is often not considered in the rush to satisfy food and energy needs.

8 *Focus on improving the composition of biomass for the replacement of oil in specific chemical feedstocks*
This will ensure that biomass can replace oil for non-transport applications.

9 *Where possible, use waste or co-products from food or forest production as a source of biomass for biofuels*
These waste products probably have the lowest environmental footprint of any source of biomass.

10 *Promote the development of alternative energy systems (not based upon carbon-based fuels) for transportation in the longer term*
This strategy will ensure that in the long term we move to the most environmentally desirable sources of energy and eliminate competition with food and biodiversity.

11 *Document biodiversity by establishing information technology systems and develop repositories of DNA and DNA sequence data (collected using emerging technologies)*
This technology will enhance the management of biodiversity and genetic resources for agricultural species.

12 *Develop second generation biofuel molecules (beyond ethanol)*
This technology will improve the energy efficiency of vehicle operation and reduce the quantity of fuel required.

13 *Identify new biofuel species and domesticate these new species for energy (and other uses) using accelerated molecular assisted selection*
This will allow efficient biofuel production in a wider range of environments. The species being targeted for biofuel production at present are probably not optimal. Most were originally domesticated for food or other uses. The full range of plant genetic resources needs to be considered.

14 *Research co-products to improve the viability of bioenergy production from plants*
This research will improve the economics of biofuel production.

15 *Encourage discussion and policy development on sustainable population growth*
Population control and reduced environmental footprint per person remains a key objective to improve our chances of a sustainable future.

16 *Ensure the features of the local or regional environment are understood and considered in finding and applying solutions for a sustainable future*
Global solutions will not always work locally. Local or indigenous knowledge needs to be recorded and communicated. The idea that there is only one approach needs to be resisted with support for the identification of locally relevant solutions to sustainability.

17 *Conduct global agricultural research and development to support food security*
We need to support the technology required to achieve a sustainable future. More resources are required to support plant genetic resource characterization and conservation. Plant breeding at all levels (from molecular tool development and application to field assessment) needs more public support. Genetic solutions to greater nutrient and water efficiency need to be complemented with continued investment in management techniques and decision-making tools. The underlying scientific skills required for this effort seem to be in decline internationally and need to be maintained and expanded.

18 *Support* ex situ *(not in the wild) collections of plant diversity at a regional, national and international level*
Plant genetic resources stored in this way assist in ensuring food security and biodiversity conservation.

19 *Actively manage populations of rare or threatened plants*
They will not always survive without intervention.

20 *Encourage responsible human diets*
Consumption of too much food and the wrong food is a major and growing problem, resulting in growing obesity in human populations.

Reduced food consumption per capita in societies with very high consumption rates would greatly ease pressure on land, water and other resources, and improve human health.

21 *Research and encourage adoption of sustainable agricultural production systems*
Knowledge of approaches to increased sustainability need to be communicated to rural communities worldwide. For example, technologies to sustain soil nutritional status and water supply and quality, while continuing to maintain agricultural productivity, need to continue to be refined and adopted.

22 *Research and promote techniques to support on-farm conservation of biodiversity*
On-farm strategies to support biodiversity with cultivated fields and in the adjoining areas need to be identified and communicated worldwide. These strategies may include the reservation of some parts of farms for nature conservation, creating refuges for plants and animals with potential benefits to productivity within the crop.

23 *Look for genetic or biological rather than chemical solutions to pests and diseases*
These approaches will deliver better environmental outcomes, reduce costs of production and deliver more sustainable high crop yields.

24 *Domesticate a wider range of species for use in food production*
Diversified food production will contribute to food security and provide a wider range of options better adapted to specific local environments.

25 *Reduce greenhouse gas emissions*
Climate change threatens food supply and biodiversity, both directly and indirectly through increased demand for land for food.

References and Further Reading

The following list includes publications cited in this book and others that provide useful further information on the topics discussed.

Adams, W. (2004) *Future Nature: A Vision for Conservation*, Earthscan, London, p256

Ahmann, D. and Dorgan, J. R. (2007) 'Bioengineering for pollution prevention through development of biobased energy and materials', State of the science report, EPA/600/R-07/028, US Environmental Protection Agency, Office of Research and Development, National Centre for Environmental Research, Washington DC, p196 – report on potential for plants to replace oil

Allendorf, F. W. and Luikart, G. (2007) *Conservation and the Genetics of Populations*, Blackwell, Oxford, p642 – an introduction to population genetics for conservation

Angiosperm Phylogeny Group (2003) 'An update of the Angiosperm Phylogeny Group classification of the orders and families of flowering plants: APG II', *Botanical Journal of the Linnaean Society*, vol 141, pp39–436

Angiosperm Phylogeny Group II (nd) www.mobot.org/MOBOT/research/APweb/, accessed 20 January 2009

Atsumi, S., Hanai, T. and Liao, J. C. (2008) 'Non-fermentative pathways for synthesis of branched-chain higher alcohols as biofuels', *Nature*, vol 451, pp86–89

Battisti, D. S. and Naylor, R. L. (2009) 'Historical warnings of future food insecurity with unprecedented seasonal heat', *Science*, vol 323, pp240–244

Beierkunlein, C. and Jentsch, A. (2005) 'Ecological importance of species diversity', in Henry, R. J. (ed.), *Plant Diversity and Evolution*, CABI, Wallingford, pp249–285

Biospectrum Asia (2008) www.biospectrumasia.com/Content/231208OTH8097.asp?nl=6_172419_Dec23, accessed 1 January 2009

Blainey, G. (2004) *A Very Short History of the World*, Penguin, Melbourne – provides useful perspectives on the roles of plants and agriculture in human history

Blakeney, A. B., Harris, P. J., Henry, R. J. and Stone, B. A. (1983) 'A simple and rapid preparation of alditol acetates for monosaccharide analysis', *Carbohydrate Research*, vol 113, pp291–299

Borlaug, N. E. (2004) 'Feeding a world of 10 billion people: Our 21st century challenge', in Scanes, C. G. and Miranowski, J. A. (eds), *Perspectives in World Food and Agriculture* 2004, Iowa State Press, Ames, IA

BP Statistical Review of World Energy (2008) http://www.bp.com/productlanding.do?categoryId=6929&contentId=7044622, accessed 14 April 2009 – a useful source of data on energy use

Bradbury, L. M. T., Fitzgerald, T. L., Henry, R. J., Jin, Q. and Waters, D. L. E. (2005a) 'The gene for fragrance in rice', *Plant Biotechnology Journal*, vol 3, pp363–370

Bradbury, L. M. T., Henry, R. J., Jin, Q. and Waters, D. L. E. (2005b) 'A Perfect Marker for Fragrance Genotyping in Rice', *Molecular Breeding*, vol 16, pp279–283

Bradbury, L. M. T., Henry, R. J. and Waters, D. L. E. (2008) 'Flavour development in rice', in Havin-Frenkel, D. and Belanger, F. C. (eds), *Biotechnology in Flavour Production*, Blackwell Publishing, Oxford, pp130–146

Bundock, P. C. and Henry, R. J. (2004) 'Single nucleotide polymorphism, haplotype diversity and recombination in the isa gene of barley', *Theoretical and Applied Genetics*, vol 109, pp543–551

Burke, J. M. (2008) *The genetic basis of crop evolution: Insights from sunflower Harlan II: Biodiversity in Agriculture: Domestication, Evolution & Sustainability*, University of California, Davis, p7

Callow, J. A., Ford-Lloyd, B. V. and Newbury, H. J. (1997) Biotechnology and plant genetic resources: Conservation and use, CABI, Wallingford, p308 – review of plant genetic resource issues

Chandel, A. K., Ravinder Rudravaram, C. E. S., Lakshmi Narasu, M., Rao, V. and Ravindra, P. (2007) 'Economics and environmental impact of bioethanol production technologies: an appraisal', *Biotechnology and Molecular Biology Review*, vol 2, pp14–32

Charles, M. B., Ryan, R., Ryan, N. and Oloruntoba, R. (2007) 'Public policy and biofuels: The way forward', *Energy Policy*, vol 35, pp5737–5746

Chase, M. (2005) Relationships between the flowering plant families, in Henry, R. J. (ed.), *Plant diversity and evolution*, CABI, pp7–23

Chase M. et al, APG (2003) 'An update of the Angiosperm Phylogeny Group classification for the orders and families of flowering plants', *Botanical Journal of the Linnean Society*, vol 141, pp399–466

Chu, Y., Faustinelli, P., Ramos, M. L., Hajduch, M., Stevenson, S., Thelen, J. J., Maleki, S. J., Cheng, H. and Ozias-Akins, P. (2008) 'Reduction of IgE binding and nonpromotion of *Aspergillus flavus* fungal growth by simultaneously silencing Ara h2 and Ara h6 in peanut', *Journal of Agricultural and Food Chemistry*, vol 56, pp11225–11233

Cleveland, D. A. and Soleri, D. (eds) (2002) Farmers, scientists and plant breeding. Integrating knowledge and practice, CABI, Wallingford, p338 – an introduction to the inclusion of farmers in plant breeding

Cline, W. R. (2007) Global warming and agriculture. Impact estimates by country. Centre for Global Development, Peterson Institute for International Economics, Washington – analysis of potential regional impacts of global warming

Computing Research, US Department of Energy (www.science.doe.gov) – a systems biology analysis of bioenergy options

Cooper, H. D., Spillane, C. and Hodgkin, T. (eds) (2001) Broadening the genetic base of crop production, CABI, Wallingford, pp452 – a review of options for diversifying crops

Cronin, J. K., Bundock, P. C., Henry, R. J. and Nevo, E. (2007) 'Adaptive Climatic Molecular Evolution in Wild Barley at the ISA Defense Locus', *Proceedings of the National Academy of Science USA*, vol 104, pp2773–2778

Cross, M., Waters, D., Lee, L. S. and Henry, R. J. (2008) 'Endonucleolytic Mutation Analysis by Internal Labeling (EMAIL)', *Electrophoresis*, vol 29, pp1291–1301

Danielsen, F., Beukema, H., Burgess, N. D., Parish, F., Bruel, C. A., Donald, P. F., Murdiyarso, D., Phalan, B., Reijinders, L., Struebig, M. and Fitzherbert, E. B. (2008) 'Biofuel plantations on forested lands: Double jeopardy for biodiversity and climate', *Conservation Biology DOI*: 10.1111/J 1523–1739 2008 01096

Darbyshire, B., Henry, R. J., Melhuish, F. H. and Hewett, R. K. (1979) 'Diurnal variations in non-structural carbohydrates, leaf extension and leaf cavity carbon dioxide concentrations in *Allium cepa* L.', *Journal of Experimental Botany*, vol 30, pp109–118

De Vicente, M. C. and Andersson, M. S. (2006) DNA banks providing novel options for genebanks. IPGRI, Rome, p7 – introduction to DNA banking

Diamond, J. (1997) *Guns, germs and steel*, Vintage, London, p480

Diamond, J. (2005) Collapse. How societies choose to fail or succeed, Penguin, New York – analysis of the history of failed societies

Dillon, S. L., Shapter, F. M., Henry, R. J., Cordeiro, G., Izquierdo, Lee, L. S. (2007) 'Domestication to Crop Improvement: Genetic Resources for Sorghum and Saccharum (Andropogoneae)', *Annals of Botany*, vol 100, pp975–989

Dinar, A., Hassan, R., Mendelsohn, R., Benhin, J. et al (2008) *Climate Change and Agriculture in Africa*, Earthscan, p206

Doctorow, C. (2008) 'Welcome to the Petacentre', *Nature*, vol 455, pp16–21

DOE Genomics: GTL Roadmap Systems biology for energy and environment (2005) Office of Biological and Environmental Research and Office of Advanced Scientific

Dow, K. and Downing, T. E. (2006) *The Atlas of Climate Change: Mapping the World's Greatest Challenge*, Earthscan, London/University of California Press, p112

Ehrlich, P. R. (1971) *The Population Bomb*, Pan/Ballantine, London

FAO (2005) Summary of World Food and Agricultural Statistics 2005, Food and Agriculture Organization of the United Nations, Rome 2005

FAO (2007) www.fao.org/statistics/yearbook/vol_1_2/world_profile.pdf, accessed 23 May 2008

FAO (2008a) www.fao.org/news/story/en/item/8836/icode/, accessed 16 December 2008

FAO (2008b) www.fao.org.ag/agl/agll/terastat/ accessed 13 October 2008

Fargione, J., Hill, J., Tilma, D., Polasky, S. and Hawthorne, P. (2008) 'Land clearing and the biofuel carbon debt', *Science*, vol 319, pp1235–1240 – analysis of the environmental impact of land clearing

Fedoroff, N. V. and Brown, N. M. (2004) *Mendel in the kitchen: a scientist's view of genetically modified food*, Joseph Henry Press, Washington – overview of the issue of genetic modification of food

Fisher, R., Maginnis, S., Jackson, W., Barrow, E. and Jeanrenaud, S. (2009) *Linking Conservation and Poverty Reduction Landscapes, People and Power*. Earthscan, London – shows links between poverty reduction and conservation

Flade, M., Schmidt, H. and Werner, A. (2006) *Nature Conservation in Agricultural Ecosystems*, Quelle & Meyer, Verlag, p706

Flannery, T. (2005) *The Weather Makers*, The Text Publishing Company, Melbourne, p332

Food and Agriculture Organization of the United Nations (2005) Summary of World Food and Agricultural Statistics, Rome – source of food production data

Frow, E., Ingram, D., Powell, W., Steer, D., Vogel, J. and Yearly, S. (2009) 'The politics of plants', *Food Security*, vol 1, pp17–23

Fry, C. (2008) *The Impact of Climate Change: The World's Greatest Challenge in the Ttwenty-first Century*, New Holland, London – introduction to climate change

Fu, Y-B. (2008) Molecular inference of flax domestication. Harlan II: Biodiversity in Agriculture: Domestication, Evolution & Sustainability, University of California, Davis, p25

Fuller, D. (2007) 'Contrasting patterns of crop domestication: Recent archaeobotanical insights from the old world', *Annals of Botany*, vol 100, pp903–924

Fuller, D. Q., Harvey, E. and Qin, L. (2007) 'Presumed domestication? Evidence for wild rice cultivation and domestication in the fifth millennium BC in the lower Yangtze region', *Antiquity*, vol 81, pp316–331

Furtado, A., Henry, R. J., Scott, K. J. and Meech, S. B. (2003) 'The Promoter of the asi Gene Directs Expression in the Maternal Tissues of the Seed in Transgenic Barley', *Plant Molecular Biology*, vol 52, pp787–799

Gerbens-Leenes, P. W., Hoekstra, A. Y., van der Meer, H. (2008) The water footprint of bioenergy. Global water use for bio-ethanol, biodiesel, heat and electricity, www.waterfootprint.org/Reports/Report34-WaterFootprint-of-Bioenergy, accessed 17 November 2008 – analysis of water use in bioenergy production

Gibson, D. G., Benders, G. A., Axelrod, K. C., Zaveri, J., Algire, M. A., Moodie, M., Montague, M. G., Venter, J. C., Smith, H. O. and Hutchison, C. A. (2008) 'One step assembly in yeast of 25 overlapping DNA fragments to form a complete synthetic *Mycoplasma genitalium* genome', *Proceedings of the National Academy of Sciences*, US, vol 105, pp20404–20409

Goddard Institute for Space Studies, NASA (2008) www.data.giss.nasa.gov/gistemp/, accessed 30 September 2008

Gravely, B. R. (2008) 'Power sequencing', *Nature*, vol 453, pp1197–1198

Guarino, L., Ramanatha Rao, V. and Reid, R. (1995) Collecting plant genetic diversity technical guidelines, CABI, Wallingford, p748 – a technical guide to collecting plant genetic resources

Hall, D. O., Scurlock, J. M. O., Bilhar-Nordenkampf, H. R., Leegood, R. C. and Long, S. P. (1993) *Photosynthesis and Production in a Changing Environment: a Field and Laboratory Manual*, Chapman & Hall, London, p464 – an introduction to photosynthesis

Hampshire, vol 3: Advanced Genomics, in press. – review of opportunities for genomics to facilitate the development of new plants

Harris, P. J., Henry, R. J., Blakeney, A. B. and Stone, B. A. (1984) 'An improved procedure for the methylation analysis of oligosaccharides and polysaccharides', *Carbohydrate Research*, vol 127, pp59–73

Harris, P. J. (2005) Diversity in plant cell walls, in Plant diversity and evolution, Henry, R. J. (ed.), CABI Publishing, pp201–227

Hartati, S., Sudarmonowati, E., Park, Y. W., Kaku, T., Kaida, R., Baba, K. and Hayashi, T. (2008) 'Overexpression of poplar cellulose accelerates growth and disturbs the closing movements of leaves in Senagon', *Plant Physiology*, vol 147, pp552–561

Haughton, A., Bond, A. J., Lovett, A. A., Dockerty, T., Sünnenberg, G., Clark, S. J., Bohan, D. A., Sage, R. B., Mallot, M. D., Mallott, V. E., Cunningham, M. D., Riche, A. B., Sheild, I. F., Finch, J. W., Turner, M. M. and Karp, A. (2009) 'A novel, integrated approach to assessing social, economic and environmental implications of changing rural land-use: a case study of perennial biomass crops', *Journal of Applied Ecology*, vol 46, pp315–322

Havkin-Frenkel, D. and Belanger, F. C. (2008) *Biotechnology of Flavor Production*, Blackwell, Oxford, p214 – an introduction to the control of flavor in novel food development

Heaton, E., Voigt, T. and Long, S. P. (2004) 'A quantitative review comparing the yields of two candidate C_4 perennial biomass crops in relation to nitrogen, temperature and water', *Biomass and Bioenergy*, vol 27, pp21–30

Held, M. A., Penning, B., Brandt, A. S., Kessans, S. A., Yong, W., Scofield, S. R. and Carpita, N. C. (2008) 'Small-interfering RNAs from natural antisense transcripts derived from a cellulose synthase gene modulate cell wall biosynthesis in barley', *Proceedings of the National Academy of Sciences USA*, vol 105, pp20534–20539

Henry, R. J. (1997) *Practical Applications of Plant Molecular Biology*, Chapman & Hall, London – introduction to plant molecular biology

Henry, R. J. (2001a) 'Exploiting cereal genetic resources', *Advances in Botanical Research,* Academic Press, vol 34, pp133–138

Henry, R. J. (ed.) (2001) Plant genotyping. The DNA fingerprinting of plants, CABI – overview of DNA analysis in plants

Henry, R. J. (ed.) (2005) Plant diversity and evolution, CABI – reviews the diversity of plants at the genetic and phenotypic level

Henry, R. J. (2005a) 'Conserving genetic diversity in plants of environmental, social or economic importance', in Henry, R. J. (ed.), *Plant Diversity and Evolution*, CABI Publishing, Wallingford, pp317–325

Henry, R. J. (2005b) 'Importance of plant biodiversity', in Henry, R. J. (ed.), *Plant Diversity and Evolution*, CABI Publishing, Wallingford, pp1–5

Henry, R. J. (ed.) (2006) *Plant Conservation Genetics*, Haworth Press, Binghamton, p180 – overview of plant conservation genetics

Henry, R. J. (ed.) (2008) *Plant Genotyping II SNP Technology*, CABI, Wallingford, UK

Henry, R. J. (2009) 'Genomics in the New Millennium', in Kole, C. and Abbott, A. G. (eds), *Principles and Practices of Plant Genomics*, Science Publishers, Inc., New York

Henry, R. J. (2010) 'Evaluation of plant biomass resources available for replacement of fossil oil', *Plant Biotechnology Journal*, in press.

Henry, R. J. and Harris, P. J. (1997) 'Molecular distinction between Monocotyledons and Dicotyledons: more than a simple dichotomy', *Plant Molecular Biology Reporter*, vol 15, pp216–218

Henry, R. J. and Kettlewell, P. S. (1996) *Cereal Grain Quality*, Chapman & Hall, London, p488 – introduction to the food uses of the major cereal crops

Henry, R. J. and Oono, K. (1991) 'Amplification of a GC-rich sequence from barley by a two-step polymerase chain reaction in glycerol', *Plant Molecular Biology Reporter*, vol 9, pp139–144

Heywood, V. H. (1978) *Flowering plants of the world*, Oxford University Press, Oxford, p336

Heywood, V. H., Brummitt, R. K., Culham, A. and Seberg, O. (2007) *Flowering Plants Families of the World*, Kew Publishing – descriptions of the flowering plants

Hill, J., Nelson, E., Tilman, D., Polasky, S. and Tiffany, D. (2006) 'Environmental, economic and energetic costs and benefits of biodiesel and ethanol biofuels', *Proceedings of the National Academy of Sciences*, USA, vol 103, 11206–11210

Hill, K. (2005) Diversity and evolution of gymnosperms, in Henry, R. J. (ed.), *Plant Diversity and Evolution*, CABI, pp25–44

Hirsch, R. L. (2007) Peaking of world oil production: Recent forecasts, www.worldoil.com/MAGAZINE_DETAIL.asp?ART_ID=3163&MONTH_YEAR=Apr_2007, accessed 23 May 2008

Hoegh-Guldberg, O., Hughes, L., McIntyre, L. S., Lindenmayer, D. B., Parmesan, C., Possingham, H. P. and Thomaset, C. D. (2008) 'Assisted colonization and rapid climate change', *Science*, vol 321, pp345–346

Huber, G. W., Iborra, S. and Corma, A. (2006) 'Synthesis of transportation fuels from Biomass: chemistry, catalysis, and engineering', *Chemical Reviews*, vol 106, pp4044–4098 – review of biomass conversion chemistries

IEA Bioenergy (2007) Potential contribution of bioenergy to the world's future energy demand, IEA Bioenergy ExCo:2007.02, www.ieabioenergy.com/MediaItem.aspx?id+5586, accessed 14 April 2009 – summary of available bioenergy resources worldwide

IEA World Energy Outlook (2008) www.worldenergyoutlook.org, accessed 26 January 2009

IFPRI (2009) www.ifpri.org/2020/focus/focus16/focus16br01.asp, accessed 14 April 2009

Imhoff, D. and Baumgartner, J. A. (2006) *Farming and the fate of wild nature: Essays on conservation-based agriculture*, University of California Press, p264

International Energy Agency (2008) *Key World Energy Statistics 2008*, Paris

IUCN red list of threatened species (2007) www.iucnredlist.org/

Karp, A., Haughton, A. J., Bohan, D. A., Lovett, A. A., Bond, A. J., Dockerty, T. et al (2009) 'Perennial energy crops: Implications and potential', in Winter, M. and Lobley, M. (eds), *What is land for? The food, fuel and climate change debate*, Earthscan, London

Keeling, C. D. and Whorf, T. P. (2005) 'Atmospheric CO_2 records from sites in the SIO air sampling network', in *Trends: A compendium of data on global ch*ange, Carbon Dioxide Information Analysis Centre, Oak Ridge National Laboratory, US Department of Energy, Oak Ridge, Tenn., USA http://cdiac.ornl.gov/ftp/trends/co2/maunaloa.co2, accessed 27 August 2008

Kovach, M. J. and McCouch, S. R. (2008) 'Leveraging natural diversity: back through the bottleneck', *Current Opinion in Plant Biology*, vol 11, pp193–200

Kovach, M. J., Sweeney, M. T. and McCouch, S. R. (2007) 'New insights into the history of rice domestication', *Trends in Genetics*, vol 23, pp578–587

Lang, T. and Heasman, M. (2004) *Food Wars*, Earthscan, London, p365

Leigh, J., Boden, R. and Briggs, J. (1984) *Extinct and Endangered Plants of Australia*, Macmillan, Melbourne, p369

Li, X., Weng, J.-K. and Chapple, C. (2008) 'Improvement of biomass through lignin modification', *Plant Journal*, vol 54, pp569–581

Lister, R., O'Malley, R. C., Toni-Filippini, T., Gregory, B. D., Berry, C. C., Millar, A. H. and Ecker, J. R. (2008) 'Highly integrated single-base resolution maps of the epigenome in Aribidopsis', *Cell*, vol 133, pp523–536

Lovejoy, T. E. (2006) *Climate Change and Biodiversity*, Earthscan

Luthi, D., Le Floch, M., Bereiter, B., Blunier, T., Barnola, J.-M., Siegenthaler, U., Raymond, D., Jouzel, J., Fischer, H., Kawamur, K. and Stocker, T. F. (2008) 'High-resolution carbon dioxide concentration record 650,000–800,000 years before present', *Nature*, vol 453, pp379–382

Lynd, L. R., Weimer, P. J., van Zyl, W. H. and Pretorius, I. S. (2002) 'Microbial cellulose utilization: fundamentals and biotechnology', *Microbiology and Molecular Biology Reviews*, vol 66, pp506–577 – review of cellulose degradation

Mastepanov, M., Sigsgaard, C., Dlugokencky, E. J., Houweling, S., Strom, L., Tramstorf, M. P. and Christensen, T. R. (2008) 'Large tundra methane burst during onset of freezing', *Nature*, vol 456, pp628–630

Maxted, N., Ford-Lloyd, B. V. and Hawkes, J. G. (1997) *Plant Conservation: the* in situ *approach*, Chapman & Hall, London p446 – review of strategies for plant conservation in the wild

McClung, R. (2008) 'Let them eat cake? One Dickens of a dilemma', *ASPB News*, vol 35, pp1–5

McDonald-Madden, E., Gordon, A., Wintle, B. A., Walker, S., Grantham, H., Carvalho, S., Bottrill, M., Joseph, L., Ponce, R., Stewart, R. and Possingham, H. P. (2008) '"True" conservation progress', *Science*, vol 323, pp43–44

McIntosh, S. R. and Henry, R. J. (2008) 'Genes of folate biosynthesis in wheat', *Journal of Cereal Science*, vol 48, pp632–638

Meadows, D. H., Meadows, D. L., Randers, J. and Behrens, W. W. (1972) *The limits to growth: a report for the Club of Rome's project on the predicament of mankind*, Earth Island, London

Meadows, D. H., Randers, J. and Meadows, D. L. (2004) *Limits to Growth: The 30-year Update*, Earthscan, London

Melbourne, B. and Hastings, A. (2008) 'Extinction risks depend strongly on factors contributing to stochasticity', *Nature*, vol 454, pp100–103

Metzger J. O. and Hüttermann, A. (2008) Sustainable global energy supply based on lignocellulosic biomass from afforestation of degraded areas. *Naturwissenschaften* doi, http://www.springerlink.com/content/21w7pq3728245414/

Millstone, E. and Lang, T. (2008) *The Atlas of Food Who Eats What, Where and Why*, 2nd edn, Earthscan, London

Motta, J. R., Serrano-Serano, M. L., Hernandez-Tores, J., Castillo-Villamizar, G. and Debouck, D. G. (2008) Domestication patterns in wild lima beans (*Phaseolus lunatus* L.) from the Americas. Harlan II: Biodiversity in Agriculture: Domestication, Evolution & Sustainability, University of California, Davis, p9

Newbury, H. J. (ed.) (2003) *Plant Molecular Breeding*, Blackwell, Oxford, p265 – an introduction to molecular plant improvement

O'Brien, N., Meizlish, M. and Hawn, A. (2008) Carbon trading and renewable energy RIRDC (www.rirdc.gov.au), Pub. No. 08/184 – discussion paper of carbon trading and renewable energy

Ochieng, J. W., Shepherd, M., Baverstock, P. R., Nikles, G., Lee, D. G. and Henry, R. J. (2008) 'Genetic variation within two sympatric spotted gum eucalypts exceeds between species variation', *Silvae Genetica*, vol 57, pp249–256

Paterson, A. H., Bowers, J. E., Bruggmann, R. and Dubchak, I. et al (2009) 'The Sorghum bicolour genome and the diversification of the grasses', *Nature*, vol 457, pp551–556

Pauly, M. and Keegstra, K. (2008) 'Cell-wall carbohydrates and their modification as a resource for biofuels', *Plant Journal*, vol 54, pp559–568

Perfecto, I., Vandermeer, J. and Wright, A. (2009) *Nature's Matrix: Linking Agriculture, Conservation and Food Sovereignty*, Earthscan, London

Potter, N. N. and Hotchkiss, J. H. (1998) *Food Science*, Springer, New York, p608 – introduction to food science

Purugganan, M. D. and Fuller, D. Q. (2009) 'The nature of selection during domestication', *Nature*, vol 457, pp843–848 – review of plant domestication

Queensland Herbarium (2008) Queensland Herbarium Achievements 2006/2007, Queensland Government, Environment Protection Agency, p61

Roberts, P. (2008) 'Carnivores like us', *Seed*, vol 16, pp60–67

Rosenberg, M. (2008) World population growth, http://geography.about.com/od/obtainpopulationdata/a/worldpopulation.htm, accessed 30 September 2008

Rosenzweig, C., Karoly, D., Vicarelli, M., Neofotis, P., Wu, Q., Casassa, G., Menzel, A., Root, T. L., Seguin, B., Tryjanowski, P., Liu, C., Rawlims, S. and Imeson, A. (2008) 'Attributing physical and biological impacts to anthropogenic climate change', *Nature*, vol 453, pp353–357

Rosillo-Calle, F., de Groot, P., Hemstock, S. L. and Woods, J. (2007) *The Biomass Handbook*, Earthscan, London – technical analysis of plant biomass

Rossetto, M., McNally, J., Henry, R. J., Hunter, J. and Matthes, M. (2001) 'Conservation genetics of an endangered rainforest tree (*Fontainea oraria* – Euphorbiaceae) and implications for closely related species', *Conservation Genetics*, vol 1, pp217–229

Rossetto, M., Jackes, B. R., Scott, K. D. and Henry, R. J. (2002) 'Is the genus *Cissus* (Vitaceae) monophyletic? Evidence from plastid and nuclear ribosomal DNA', *Systematic Botany*, vol 27, pp522–533

Rozema, J., Aerts, R. and Comelissen, H. (2006) 'Plants and climate change', *Plant Ecology*, vol 182, special issue

Rubin, E. M. (2008) 'Genomics of cellulosic biofuels', *Nature*, vol 454, pp841–845

Scanes, C. G. and Miranowski, J. A. (eds) (2004) *Perspectives in world food and agriculture 2004*, Iowa State Press, Ames, p485 – reviews global potential for food and agriculture

Scharlemann, J. P. W. and Laurance, W. F. (2008) 'How green are bio fuels?' *Science*, vol 319, pp43–44

Schiermeier, Q., Tollefson, J., Scully, T., Witze, A. and Morton, O. (2008) 'Electricity without carbon', *Nature*, vol 454, pp816–823

Schmer, M. R., Vogel, K. P., Mitchell, R. B. and Perrin, R. K. (2008) 'Net energy of cellulosic ethanol from switchgrass', *Proceedings of the National Academy of Sciences, USA*, vol 105, pp464–469

Schmidt, L. D. and Dauenhauer, P. J. (2007) 'Hybrid routes to biofuels', *Nature*, vol 447, pp914–915

Schwietzke, S., Ladisch, M., Russo, L., Kwant, K., Makinen, T., Kavalov, B., Maniatis, K. et al (2008) Analysis and identification of gaps in research for the production of second-generation liquid transportation fuels, IEA Bioenergy: T4192):2008.01 – identifies key research needs and gaps for biofuel production

Scott, P. T., Pregel, L., Chen, N., Hadler, J. S., Djordjevic, M. A. and Gresshoff, P. M. (2008) *Pongamia pinnata*: An untapped resource for the biofuels industry of the future, Bioenergy Research DOI 10.1007/s12155-008-9003-0

Simmons, B. A., Loque, D. and Blanch, H. W. (2008) 'Next-generation biomass feedstocks for biofuel', *Genome Biology*, vol 9, p242

Simpson, M. G. (2006) *Plant Systematics*, Elsevier Academic Press, Amsterdam, p590 – overview of plant systematics

Simpson, M. J. (2006) *Plant Systematics*, Elsevier, Amsterdam, p15

Smalley, J. and Blake, M. (2003) 'Sweet beginnings: stalk sugar and the domestication of maize', *Current Anthropology*, vol 44, pp675–703

Smith, B. D. (2006) 'Seed size increase as a marker of domestication in squash (*Cucurbita pepo*)', in *Documenting Domestication: New Genetic and Archaeological Paradigms*, University of California, Berkeley, pp25–31

Smith, A. M. (2008) 'Prospects for increasing starch and sucrose yields for bioethanol production', *Plant Journal*, vol 54, pp546–558

Sohmer, S. H. and Gustafson, R. (1987) *Plants and flowers of Hawaii*, University of Hawaii Press, Honolulu

Somleva, M. N., Snell, K. D., Beaulieu, J. J., Peoples, O. P., Garrison, B. R. and Patterson, N. A. (2008) 'Production of polyhydroxybutyrate in switchgrass, a value-added co-product in an important lignocellulosic biomass crop', *Plant Biotechnology Journal*, vol 6, pp663–678

Stephenson, H. (2009) Global warming: corn to the rescue? www.sciencenow.sciencemag.org/cgi/content/full/20009/115/2, accessed 21 January 2009

Stokstad, E. (2009) Farming strides towards sustainability, www.sciencenow.sciencemag.org/cgi/content/full/2009/112/4, accessed 21 January 2009

Suzuki, M. and Chatterton, N. J. (1993) *Science and Technology of Fructans*, CRC Press, Boca Raton, p369

Sweeney, M. and McCouch, S. (2007) 'The complex history of the domestication of rice', *Annals of Botany*, vol 100, pp951–957

Taylor, L. E., Dai, Z. Y., Decker, S. R. et al (2008) 'Heterologous expression of glycosyl hydrolases in planta: a new departure for biofuels', *Trends in Biotechnology*, vol 26, pp413–424

Thomas, C. D., Cameron, A., Green, R. E., Bakkenes, M., Beaumiont, L. J., Collingham, Y. C. et al (2004) 'Extinction risks from climate change', *Nature*, vol 427, pp145–148

Thompson, R. G. (2006) Documenting the presence of maize in central and south American through phytolith analysis of food residues,

Documenting Domestication. New genetic and archaeological paradigms, University of California, Berkeley, pp82–122

US Department of Energy and US Department of Agriculture (2005) Biomass as a feedstock for a bioenergy and bioproducts industry: The technical feasibility of a billion-ton annual supply, DOE/GO-102995-2135

US Department of Energy (2008) Bioenergy Research Centers. An overview of the science, DOE/SC-0104

Van Beilen, J. B. and Poirier, Y. (2008) 'Production of renewable polymers from crop plants', *Plant Journal*, vol 54, pp684–701

Vaughan, D. A., Lu, B.-R. and Tomooka, N. (2008) 'Was Asian rice (*Oryza sativa*) domesticated more than once?', *Rice*, vol 1, pp16–24

Vermerris, W. (2008a) 'Why bioenergy makes sense', in Vermerris, W. (ed.), *Genetic Improvement of Bioenergy Crops*, Springer, New York, pp3–42

Vermerris, W. (2008b) *Genetic Improvement of Bioenergy Crops*, Springer, New York, p449 – review of options for genetic improvement of current biofuel crop candidates

Von Braun, J. (2007) The world food situation. New driving forces and required actions, International Food Policy Research Institute, Washington, DOI: 10.2499/0896295303

Von Braun, J. (2008) 'The food crisis isn't over', *Nature*, vol 56, p701

Waters, D. L. E. and Henry, R. J. (2007) 'Genetic manipulation of starch properties in plants: Patents 2001–2006', *Recent Patents on Biotechnology*, vol 1, pp252–259

Waters, D. L. E., Henry, R. J., Reinke, R. F. and Fitzgerald, M. A. (2006) 'Gelatinization temperature of rice explained by polymorphisms in starch synthase', *Plant Biotechnology Journal*, vol 4, pp115–122

Watson, L. and Dallwitz, M. J. (1992 onwards) The families of flowering plants: descriptions, illustrations, identification, and information retrieval. Version: 25 November 2008, http://delta-intkey.com, accessed 20 January 2009

WHO (2008) Global and regional food consumption patterns and trends, www.who.int/nutrition/topic/3-foodconsumption, accessed 7 October 2008

World Meteorological Organization (2008) www.wmo.int/pages/mediacentre/press_releases/pr_833_en.html, accessed 12 April 2009

Worldwatch Institute (2006) *Biofuels for Transport: Global Potential and Implications for Energy and Agriculture*, Earthscan, London – reviews the potential for biofuel production

Zeder, M. A., Bradley, D. G., Emshwiller, E. and Smith B. D. (eds) (2006) *Documenting Domestication: New Genetic and Archaeological Paradigms*, University of California Press, Berkeley – overview of domestication

Index

Acacia 67
adzuki beans 12, 133
affordable food 118
Africa 43
Africa Rice Centre 34
agricultural research 161
albedo 122
alcohol 54
alfalfa 12
algae 75
alkanes 61, 64
allergy 153
Allocasuarina 72
Amazon 84, 119
amino acids 24
ancient grain 128
angiosperm 141, 142
Angiosperm Phylogeny Group 95
animal feed 5
anti-nutritional factors 12
apple 8, 30, 54
apricots 54
aquatic plants 101
arabinoxylans 56
arable land 2, 83
Arachis hypogaea 130
archaeological records 129
aroma 17
arrowroot 8
Asia 44
asparagus 8
aubergine 8
Aus 131
Australia 44, 85
Avena sativa 10
avocado 8
bagels 11
balady 11
bamboo 8
banana 8, 12, 130
Banksia conferta 111
barabari 11
barley 8, 9, 10, 30, 45, 67, 121, 126, 130, 159
basmati 17, 131
bay leaves 8
beans 8
beans (green) 12
beef 7, 19
beer 5, 13
beta-glucan 56
beverages 5, 13
biochemial conversion 59
biodiesel 54, 62
biodiversity 100, 157, 162
biodiversity in cultivation 102
bioenergy 50
biofuel 53, 145
biofuel production 51
biomass transportation 76
biorefinery 66, 76
Bioversity International 34
black gram 12
borlotti bean 12
botanic gardens 102, 103
bottle gourd 126
Brazil 83, 85, 119
breads 11
breakfast foods 11
broad bean 12
Bryophytes 141
bulk density 66, 75
1-butanol 63

butanol 62, 64

cabbage 8
calorie intake 10
camelina 67, 72
Camelina sativa 72
camphor laurel 118
Canada 45
canola 24, 66, 67, 74
capsicum 30
carbohydrate 11, 23, 24, 54
carbon balance 77
carbon dioxide 37, 38
carbon storage 155
carbon trading 116
cardamom 8
caribbean pine 150
carob 12
carpets 65
carrot 8, 30
cashew 8
cassava 8, 13, 30, 121, 130, 139
Cassia 67
castor oil 24, 67, 74
casuarina 67, 72
cats 129
cauliflower 8
CBOL 97
celery 8
cell wall 24
cellulose 24, 56
cellulosic biomass 86
Centre for International Forestry
 Research 34
cereal 5, 9, 45, 55
cereal breeding 32
Cerrado 84, 119
CGIAR 33, 34
chapatti 11
charcoal wood 23
chemical conversion 60
chemical feedstocks 51, 53
cherry 8, 54
chestnut 8
chickpea 12, 30, 127
chicory 8
Chikusichloa 28
China 2, 19
CIAT 34
CIFOR 34
CIMMYT 34
Cinnamomum camphor 118
cinnamon 8
CIP 34
Cissus 133
Cissus antarctica 27
citrus 12
climate change 37, 40, 109
clover 8
cloves 8
clubmosses 141
CO2 37
coal 47
coastal fontainea 94
coconut 8, 24, 99
Cocus nucifera 99
coffee 5, 8
common bean 130
competition for land 121
competition from weeds 98
composition of plants 23
Conifers 141
conservation genetics 96
Consortium for the Barcode of Life 97
Consultative Group on International
 Agricultural Research 33, 34
cooking requirements 17
copra 8
co-product 66
Corymbia variegata 98
cosmetics 5
cotton 5, 130
cowpea 30
cows 13, 86, 127
critically endangered 94
croissant 11
crop residues 86, 120
crumpets 11
cryopreservation 106
cucumber 8
Cucurbita pepo 130
currants 8
custard apple 8
cut flowers 5

Cycads 141

danish 11
dates 8
Davidsonia 135
Davidsonia jerseyana 136
Davidsonia johnsonii 136
desertification 36
diesel 62
Diesel tree 67
dietary fibre 24
dill 8
2, 5-dimethylfuran 60
dimethylfuran 62
Dioscorea 13
Dioscorea rotundata 130
diversity within species 95
DMF 60
DNA analysis 29, 96
DNA banks 25, 29, 31, 96, 103, 104
DNA fingerprinting 151
DNA sequencing 80, 141
DNA sequencing technology 79
DNA technology 32
dogs 129
domestication 125, 132, 140

ecosystem services 101
eggs 9, 14
El Nino 46
elderberry 8
electricity 5, 53
electricity generation 50
Eluesine corocana 10
endangered 94
energy crops 160
energy resources 47
energycane 69
English muffin 11
environmental pollutants 98
environmental sustainability 158
enzymes 59
erucic acid 12
ethanol 5, 62, 63
eucalypt 67, 71, 73, 149
eucalyptus 59
eudicots 95
Europe 44
evolutionary processes 97
ex situ 25
extinct 94
extinct in the wild 94

Faba bean 12, 30
fast pyrolysis 60
fats 24
fatty acids 65
fennel 8
fermentation 62
ferns 141
Fertile Crescent 1, 15, 126, 127
fertilizer 51
ferulic acid 56
fibre 5
fig 8
fires 112
firewood 5
first generation 68, 75
first generation technology 63
fish 14, 19
Fisher-Tropsch 62
flax 130, 134
flowering plants 7, 95, 141
fodder 5
folate 35
Fontainea oraria 94
food 5
food and energy prices 117
food chain 91
food consumption 21, 22
food crops 159
food deficiencies 36
food demand 87, 89
food from animals 13
food ingredients 66
food labelling 19
food miles 19
food prices 34, 81
food processing 150, 152
food production 7
food security 161
food supply 89
forestry 119
fossil fuels 47

fragrance 17
fructans 55
fruit 5, 8, 9, 12
fuel 5
fuel wood 23

garden plants 5
garlic 8
gas 47
gasification 60
gasoline 62
gene pool 25
genetic code 80
genetic resources 24
genomics 31, 154
geothermal 50
geothermal energy 50
germplasm 30
Giant cane 67
ginger 8
Ginkgo 141
Global Crop Diversity Trust 104
Global Seed Vault 104
global temperature 39, 40
glucose 23, 63
glucosinolates 12
Gnetales 141
goat 13, 127
gooseberries 8
Gossypium 130
Gossypium arboretum 130
Gossypium barbadense 130
Gossypium herbaceum 130
Gossypium hirsutum 130
grain legumes 11
grape 8, 12, 13, 27, 30, 130
grazing 98
green revolution 32, 88, 143
greenhouse 110
greenhouse gas balance 117
greenhouse gas emissions 53, 162
greenhouse gases 5, 38
gymnosperm 142

haricot beans 12
harvest index 32, 66
Hawaii 99
heat stable amylase 55
Helianthus annuus 130, 132
hemp 5
herbaria 103
heterosis 149
Hicksbeachia pinnatifolia 135
higher level diversity 95
Higher plants 141
HMF 60
hops 8
Hordeum spontaneum 15, 130
Hordeum vulgare 10, 130
hornworts 141
Horsetails 141
human diets 9, 161
human energy consumption 48
human food preferences 14
human population 19, 34
hybrid plants 32
hybrid poplars 71
hybrid vigour 149
hybrids 29
hydro 50
hydrogen 51
hydrological impact 117
5-hydroxymethylfurfural 60
Hygroryza 28

ICARDA 34, 126
ICRAF 34
ICRISAT 34
IFPRI 34
in situ 25
in situ conservation 106
India 2, 19
Indica 131
International Center for Agricultural Research in the Dry Areas 34, 126
International Centre for Agricultural Research in the Semi-Arid Dry Tropics 34
International Centre for Tropical Agriculture 34
International Centre for Wheat and Maize Improvement 34
International Food Policy Research Institute 34

International Potato Research Centre 34
International Rice Research Institute 34
International Union for Conservation of Nature and Natural Resources 93
Internet 156
IRRI 34
IUCN Red List 93

Japan 115
Japonica 131
Jarvonica 131
jatropha 67, 74, 121
Jatropha curcas 74
Jerusalem artichoke 8
jet fuel 62
Jojoba 67

kava 8
kidney beans 12

Lagenaria siceraria 126
land availability 82
land clearing 106
land use 119
lantana 99
Lantana camara 99
LCA 116
leek 8
Leersia 28
lemon 8
lentil 12, 30, 127
lettuce 8
life cycle assessment 86
lignin 24, 56, 61
lillipilly 8
Lima bean 130, 133, 134
lime 8
Linum angustifolium 130
Linum usitatissimum 130, 134
liverworts 141
living collections 25, 103
living standards 2
loss of plant diversity 109
lucerne 8
lupin 12, 30
Luziola 28
lychee 8
Lycophytes 141

macadamia 8, 25, 26
Macadamia integrifolia 134
Macadamia tetraphyla 134
magnolids 95
mahogany 103
maize 8, 9, 10, 14, 16, 30, 62, 66, 67, 121, 130, 159
Maltebrunia 28
mango 8
Manihot esculenta 130
maple sugar 8
Mayan society 147
meat 3, 9, 34
mechanized harvesting 70
medicine 5
melaleuca 73
melon 8
Meso-America 16
2-methyl-1-butanol 63
3-methyl-1-butanol 63
Mexico 85, 99
milk 9
millet 8, 10, 30
Miscanthus 67, 70, 116, 139
Miscanthus X giganteus 70
modern cultivars 132
molasses 69
monocotyledons 58, 95
mosses 141
mulberries 8
multi-purpose crops 85, 155
mung bean 12
Musa 130
Musa acuminata 130
Musa balbisiana 130
mustard 8
mutation breeding 31

N use efficiency 66
N_2O 38
naan 11
nanotechnology 80
natural gas 48
nature conservation 123
new technology 122

newspaper 59
nitrogen fixation 11
nitrous oxide 38
non food uses 23
non structural carbohydrates 54, 55
non-cellulosic polysaccharides 24
noodles 11
North America 44
nuclear power 48
nutmeg 8
nutrient run-off 117
nutritional value 33

oat 10, 30
obesity 21, 36
oceans (waves and tides) 50
oil 11, 47, 61
oil consumption 48, 49
oil palm 12, 24, 67
oil prices 82, 117
oilcrops 72
oilseed rape 8
oilseeds 5, 9, 11, 12
Olea europaea 130
olive 8, 24, 67, 130
onion 8, 30
orange 8
orchids 105
ornamentals 105
Oryza 28
Oryza barthii 130
Oryza glaberrima 10, 130, 131
Oryza rufipogon 130, 131
Oryza sativa 10, 27, 130, 131

palm oil 8, 74
pan bread 11
Panicum virgatum 70
papaw 8
paper 5, 23
Paraserianthes falcataria 79
paratha 11
parsley 8
passionfruit 8
pasta 11
pasture 5, 8, 120
p-coumeric acid 56
pea 8, 12, 30, 127
peaches 54
peanut 12, 24, 30, 130, 153
peanut (groundnut) 8
pear 8, 54
pecan 8
Pennisetum glaucum 10
pepper 8
per capita consumption 34
perfumes 5
pests and diseases 98
PHA 65
Phaseolus 130
Phaseolus lunatus 130, 134
Phaseolus vulgaris 130
2-phenylethanol 63
photosynthesis 41, 47
pig 13, 127
pigeon pea 12, 30
pine 59, 67, 139
pineapple 8
pinto bean 12
Pinus caribaea 150
Pinus elliottii 139, 150
pipelines 77
pistachio 8
pita 11
pizza crust 11
PLA 65
plant breeders 46, 149
plant carbohydrates 57
plant collectors 101
plant diversity 97, 98, 101, 113
plant domestication 2
plant genomics 151
plastic 66
plum 8, 54
policies 116
pollen storage 104
Polyhydroxyalkanoate 65
Polylactic acid 65
polysaccharides 24
polyunsaturated oil 72
Pongamia 140
Pongamia pinnata 74
pongamia tree 67, 74
poori 11

poplar 59, 67, 71, 139
population 2
population control 161
population growth 20, 118
pork 19
Porteresia 28
post harvest spoilage 152
Potamophila 28
Potamophila parviflora 27
potato 8, 13, 30, 121, 130, 139, 159
poultry 19
pre-treatments 59
pretzels 11
private plant collections 101
Prosphytochloa 28
protein 11, 12, 24
public parks 103
pulses 5, 9, 11, 12, 127
pumpkin 8
pyrolysis oil 60

rainforests 123, 104
raisin bread 11
rapeseed 121
rare plants 101
rare species 96, 101
raspberries 8
relocation of species 112
renewable energy 50
renewable energy targets 116
Rhynchoryza 28
rice 8, 9, 10, 27, 28, 30, 45, 59, 67, 121, 130, 159
Ricinus communis 74
rising sea levels 39
roots 9, 13
roti 11
round wood 23
rubber 65
ruminant animals 86
rye 10, 30, 121

Saccharum 130
Saccharum officinarum 130, 138
Saccharum robustum 138
Saccharum spontaneum 130
safflower 24, 67
saffron 8
sago 8
sake 13
salinity 36
Salix 71
sandwich buns 11
sarsaparilla 8
sawn wood 23
sea water 75
Secale cereale 10
second generation technology 63, 79
seed banks 25, 104
sheep 13, 86, 127
slash pine 150
small populations 97
soil nutrients 32
soils 83
Solanum species 130
Solanum tuberosum 130
solar power 48
sorghum 8, 9, 10, 27, 29, 30, 45, 67, 68, 121, 130, 159
Sorghum bicolor 10, 130
South America 44
soybean 8, 12, 24, 30, 67, 121
species diversity 91
squash 130
staple crops 159
starch 23, 55
steamed breads 11
strawberries 8
structural carbohydrates 56
sucrose 23, 55
sugar 5, 9, 13, 54, 59
sugar alcohols 57
sugar beet 23, 67, 121
sugarcane 8, 13, 23, 59, 64, 67, 68, 69, 119, 121, 130
sunflower 8, 12, 24, 66, 67, 130, 132
sweet potato 8, 30, 139
sweet sorghum 138
switchgrass 67, 70, 139
syngas 60

tamarind 12
targeted mutagenesis 79
taste 14

tax 116
tea 5, 8
tea tree 73
thermochemical conversion 60
tissue culture 105
tomato 8, 30
tortillas 11
toxic plants 134
traditional village garden 102
transport 49
Tree crops 71
triglycerides 65
Triticale 30
Triticum aestivum 10, 130
Triticum durum 10
Triticum monococum 130
Triticum speltoides 130
Triticum tauschii 130
trypsin 12
tubers 8, 9, 13
tumeric 8
turf grass 5

udon noodles 11
urbanization 34

vanilla 8
vegetables 5, 8, 9, 12
vetches 12
Vigna angularis 133
vitamins 35
Vitis 133
Vitis vinifera 130
vulnerable 94

walnut 8
WARDA 34
water use efficiency 66, 121
weeds 92, 99, 118
Western Australia 84
wheat 8, 9, 10, 18, 30, 45, 59, 67, 121, 126, 130, 159
white cinnamon 8
wild barley 15, 110
wild crop relatives 42
wild grape 26
wild populations 132
wild rice 27
wild wolf 129
wilderness 111
willow 59, 67, 71, 116, 139
wind 48
wind power 50
wine 5, 13
wood 71
World Agroforestry Centre 34
World Meteorological Organization 41
world population 20

xyloglucan 56

yam 8, 13, 30, 130

Zea mays 10, 130
Zea mays ssp. *Parviglumis* 130
Zizania 27, 28
Zizaniopsis 28